About this book

This workbook contains practice to support your learning

Questions split into three levels of increasing difficulty – Challenge 1, Challenge 2 and Challenge 3 – to aid progress.

Symbol to highlight questions that test problem-solving skills.

Total marks boxes for each challenge and topic.

'How am I doing?' checks for self-evaluation.

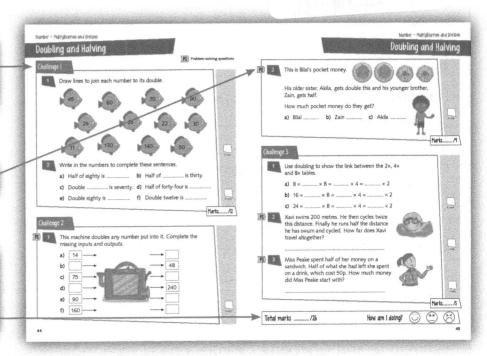

Starter test recaps skills covered in P3/P4.

Four progress tests throughout the book, allowing children to revisit the topics and test how well they have remembered the information.

Progress charts to record results and identify which areas need further revision and practice.

Answers for all the questions are included in a pull-out answer section at the back of the book.

Contents

Starter Test 4

Number – Number and Place Value

Place Value 12
Representing Numbers 14
Reading and Writing Numbers 16
10 and 100 More or Less 18
Counting in Multiples 20
Comparing and Ordering Numbers 22
Number Problems 24

Number – Addition and Subtraction

Adding and Subtracting Mentally 26
Number Bonds 28
Adding in Columns 30
Subtracting in Columns 32
Estimating and Checking Calculations 34
Addition and Subtraction Problems 36

Progress Test 1 38

Number – Multiplication and Division

Multiplication and Division Facts 42
Doubling and Halving 44
3×, 4× and 8× Tables 46
Mental Multiplication and Division 48
Multiplying 2-digit Numbers 50
Multiplication and Division Problems 52

Fractions (including Decimals)

Tenths 54
Recognising Fractions 56
Fractions of Amounts 58
Addition and Subtraction of Fractions 60
Equivalent Fractions 62
Comparing and Ordering Fractions 64
Fraction Problems 66

Progress Test 2 68

Contents

Measurement

Reading Scales 72
Comparing Measures 74
Adding and Subtracting Measures 76
Money 78
Time 80
Units of Time 82
Duration of Events 84
Perimeter of Shapes 86

Progress Test 3 88

Geometry – Properties of Shapes

2-D Shapes 92
3-D Shapes 94
Lines 96
Right Angles 98
Angles and Turns 100

Statistics

Bar Charts 102
Pictograms 104
Tables 106

Progress Test 4 108

Answers (pull-out) 113

Progress Test Charts (pull-out) 127

ACKNOWLEDGEMENTS

The author and publisher are grateful to the copyright holders for permission to use quoted materials and images.

All illustrations and images are ©Shutterstock and ©HarperCollinsPublishers.

Every effort has been made to trace copyright holders and obtain their permission for the use of copyright material. The author and publisher will gladly receive information enabling them to rectify any error or omission in subsequent editions. All facts are correct at time of going to press.

Published by Leckie
An imprint of HarperCollins Publishers
Westerhill Road, Glasgow, G64 2QT

HarperCollins Publishers
Macken House, 39/40 Mayor Street Upper, Dublin 1
D01 C9W8 Ireland

© 2024 Leckie

ISBN 9780008665883

10 9 8 7 6 5 4 3 2 1

All rights reserved. No part of this publication may be reproduced, stored in a retrieval system, or transmitted, in any form or by any means, electronic, mechanical, photocopying, recording or otherwise, without the prior permission of Collins.

British Library Cataloguing in Publication Data.

A CIP record of this book is available from the British Library.

Series Concept and Development: Michelle I'Anson
Commissioning Editor: Richard Toms
Series Editor: Charlotte Christensen
Author: Shaun Stirling
Project Manager and Editorial: Tanya Solomons
Cover Design: Sarah Duxbury
Cover Illustration: Louise Forshaw
Inside Concept Design: Ian Wrigley
Text Design and Layout: Contentra Technologies
Artwork: Collins and Contentra Technologies
Production: Natalia Rebow

Printed in United Kingdom.

PS Problem-solving questions

1. Write these numbers in digits.

    a) Sixteen _____   **b)** Sixty-one _____

    c) One hundred and eight _____   **d)** Fifty _____

4 marks

2. Put these numbers in order from lowest to highest.

lowest   [ ] [ ] [ ] [ ] [ ] [ ] [ ]   highest

1 mark

3. Fill in the missing steps in these number tracks.

    a) | 4 | 6 | | 10 | 12 | | 16 |

    b) | 18 | | 12 | 9 | | 3 | 0 |

    c) | 15 | 20 | | | 35 | 40 | |

    d) | 100 | 90 | | 70 | | | 40 |

4 marks

4. These crosses are from a 1–100 square. Fill in the missing numbers that are 1 more or less and 10 more or less.

    a)   **b)**   **c)**   **d)**

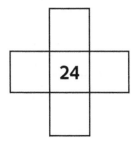 24   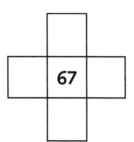 67   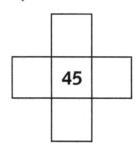 45   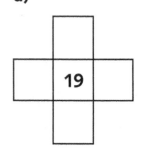 19

4 marks

5. What numbers are represented here?

a)  b)   c)

[   ]   [   ]   [   ] p

3 marks

6. Write < (less than), > (greater than) or = (equal to) to make these correct.

a) 67 [   ] 76

b) thirty-three [   ] thirteen

c) 101 [   ] 110

d) forty-four [   ] forty-four

4 marks

7. Fill in the missing addition and subtraction statements for each family of facts.

17 + 3 = 20	3 + 17 = 20	20 – 3 = 17	20 – 17 = 3
	4 + 16 = 20		20 – 16 = 4
15 + 5 = 20		20 – 5 = 15	
	6 + 14 = 20		20 – 14 = 6

6 marks

8. Fill in the missing numbers to make these correct.

a) 17 – 9 = [   ]   b) 22 – [   ] = 18   c) 9 + [   ] = 15

d) [   ] + 7 = 16   e) 12 + 15 = [   ]   f) [   ] – 11 = 9

6 marks

PS **9.** a) Laila has 26 cards in her album. She gives 7 to her friend.

How many cards does she have now? _____

b) A plant was 17 cm tall. Last week it grew 8 cm.

How tall is the plant now? _____

2 marks

5

PS Problem-solving questions

10. Look at this array of cats. What two multiplications does it show?

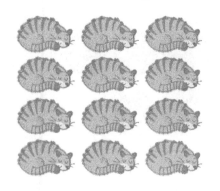

_____ × _____ = _____ and _____ × _____ = _____

2 marks

11. Complete these multiplications.

a) $9 \times 2 =$ _____   b) $3 \times 3 =$ _____   c) $8 \times 5 =$ _____

d) $10 \times 10 =$ _____   e) $7 \times 3 =$ _____   f) $5 \times 2 =$ _____

6 marks

PS **12.** Tomasz has 20p to spend. If all his coins were the same, how many coins would he have if they were all the following coins?

a) 1p coins   _____   b) 2p coins _____

c) 10p coins _____   d) 5p coins _____

4 marks

PS **13.** How many sides are there on these shapes?

a) Three squares _____   b) Five triangles _____

c) Two hexagons _____

3 marks

14. Colour the table cells. Use the same colour for each number, its half and its double. One has been done for you.

Number halved	Number	Number doubled
50	100	12
10	6	16
3	8	200
4	12	24
6	20	40

4 marks

PS **15.** Look at the chocolates below.

    a) $\frac{1}{2}$ of these chocolates have soft centres. How many is this?

    _____

    b) $\frac{1}{4}$ of these chocolates have nuts in. How many is this? _____

    c) $\frac{3}{4}$ of these chocolates are milk chocolates. How many is this?

    _____

3 marks

16. Fill in the gaps in this fraction number track.

| $\frac{3}{4}$ | 1 | | $1\frac{1}{2}$ | | 2 | $2\frac{1}{4}$ | | $2\frac{3}{4}$ |

3 marks

17. Use a ruler to measure the length and width of each rectangle.

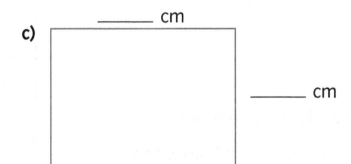

    a) _____ cm _____ cm

    b) _____ cm _____ cm

    c) _____ cm _____ cm

3 marks

18. a) Read these scales.

_____ g of flour

b) How far has the girl jumped?

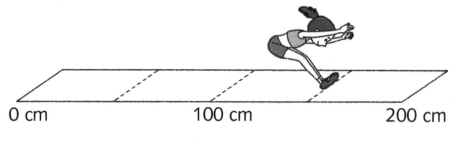

0 cm   100 cm   200 cm

_____ cm

c) What is the temperature of the water?

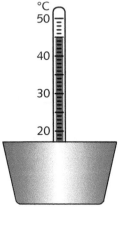

_____ °C

d) How much water is in the jug?

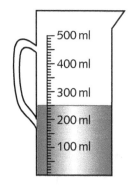

_____ ml

4 marks

19. Draw the hour and minute hands to show these times.

a) Half past 12   **b)** Quarter past 3   **c)** Quarter to 6

3 marks

PS **20.** Lena has four coins.

a) How much does she have? _____

    1 mark

b) Lena's four friends all have this amount of money too. But they all have different numbers of coins. Draw the coins they might have.

Aisling		Ryan	
Sam		Gaby	

    4 marks

PS **21.** At the school fair there is a snack bar. The following items are on sale.

20p   30p   10p

a) Fill in the gaps.

    Owen buys a drink and a cupcake, which altogether

    cost _____p. He pays with a £1 coin and gets _____p change.

    2 marks

b) Sarah buys some items and pays with a £1 coin. She gets 50p change. What might Sarah have bought? List four different possibilities.

    _____   _____

    _____   _____

    4 marks

22. Use coloured pencils to shade the hexagon(s), rectangle(s), triangle(s) and pentagon(s). Use a different colour for each type of shape.

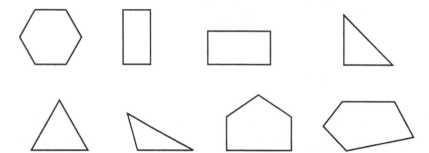

4 marks

23. Complete this table to show the properties of the 3-D shapes.

Shape	Faces	Edges	Vertices
	_____	12	_____
	5	_____	_____
	_____	9	_____

6 marks

24. This clock has only one hand.

Draw where the hand would be pointing after

a) a half turn anticlockwise

b) a quarter turn clockwise

c) three quarter turns clockwise

3 marks

25. Tanya counted car colours.

Colour	
Red	
White	
Black	
Silver	

= 2 cars

a) How many red cars did she see? _____

b) How many more white cars than black cars did she see? _____

c) Tanya counted 7 silver cars. Add them to her pictogram.

3 marks

Marks........ /96

Place Value

Challenge 1

1 Partition these numbers into tens and ones.

a) 67 = ☐ + ☐   b) 19 = ☐ + ☐

c) 88 = ☐ + ☐   d) 55 = ☐ + ☐

e) 24 = ☐ + ☐   f) 70 = ☐ + ☐

6 marks

2 What is the value of the red digit in these numbers?

a) 24 _____   b) 19 _____

c) 98 _____   d) 66 _____

4 marks

3 What number will you get if you put these numbers together?

a) Six tens and two ones   _____

b) One ten and nine ones   _____

c) Three tens and four ones _____

d) Four tens and eight ones _____

4 marks

Marks......... /14

Challenge 2

1 Partition these numbers into hundreds, tens and ones.

a) 419 = ☐ + ☐ + ☐

b) 229 = ☐ + ☐ + ☐

c) 305 = ☐ + ☐ + ☐

d) 980 = ☐ + ☐ + ☐

4 marks

Place Value

2 What is the value of the red digit in these numbers?

a) 789 _____

b) 405 _____

c) 176 _____

d) 999 _____

4 marks

Marks.......... /8

Challenge 3

1 What number will you get if you put these numbers together?

a) Eight hundreds, four tens and four ones? _____

b) Five hundreds and four ones? _____

c) One hundred, and sixteen tens? _____

d) Forty-eight tens and three ones? _____

4 marks

2 Write < (less than), > (greater than) or = (equals) between each pair to make them correct.

a) sixteen tens ☐ one hundred + six tens

b) two hundreds + four ones ☐ twenty-one tens

c) fifty + five ones ☐ fifty-four ones

d) three hundred + eight tens + two ones ☐

thirty-seven tens + twelve ones

4 marks

Marks.......... /8

Total marks .... /30

How am I doing?

Representing Numbers

PS Problem-solving questions

Challenge 1

1 What numbers are represented here?

a)

b)

c)

d)

2 How much money is in each purse?

a)

_____p

b)

_____p

4 marks

2 marks

Marks.......... /6

Challenge 2

1 What numbers are the arrows pointing to on these number lines?

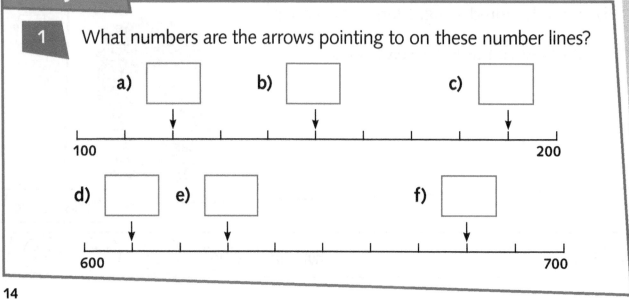

a) b) c)

100   200

d) e) f)

600   700

6 marks

14

Representing Numbers

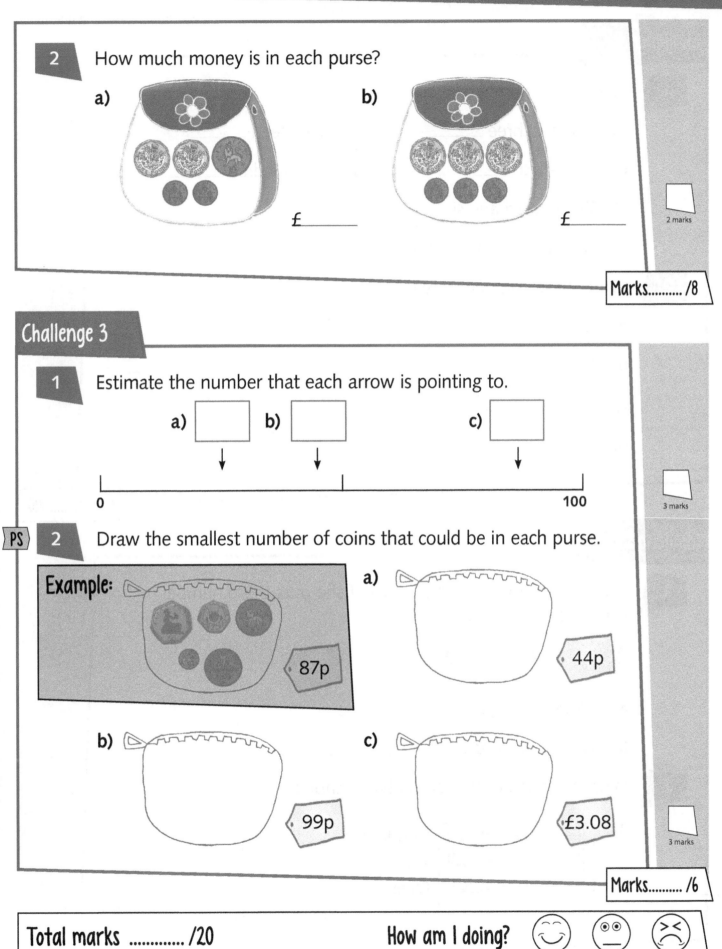

2 How much money is in each purse?

a)

b)

£_____

£_____

2 marks

Marks.........../8

Challenge 3

1 Estimate the number that each arrow is pointing to.

a) [ ]   b) [ ]   c) [ ]

0   100

3 marks

PS **2** Draw the smallest number of coins that could be in each purse.

Example:

87p

a)

44p

b)

99p

c)

£3.08

3 marks

Marks.........../6

Total marks ..../20   How am I doing? 😊 😐 😣

Reading and Writing Numbers

PS Problem-solving questions

Challenge 1

1 Write these numbers in digits.

a) Seventy-three _____

b) Seventeen _____

c) Forty-four _____

d) Two hundred _____

e) Six hundred and six _____

f) One hundred and fifty-eight _____

6 marks

2 Write these numbers in words.

a) 35 _____

b) 11 _____

c) 450 _____

d) 207 _____

4 marks

Marks......... /10

Challenge 2

1 What number will Lilly-May have if she puts these number cards together?

| 9 0 | 8 | 5 0 0 |

a) Write it in digits. _____

b) Write it in words. _____

2 marks

2 Write the number that each abacus shows.

a)

Hundreds
Tens
Ones

_____

16

Reading and Writing Numbers

b)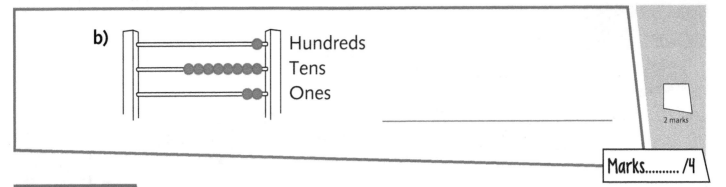

Hundreds
Tens
Ones

_____

2 marks

Marks.......... /4

Challenge 3

PS | 1 | Charley has four digit cards. | 8 | 1 | 7 | 4

First, he uses them all to make the largest 4-digit number he can.

a) Write it in digits. _____

b) Write it in words. _____

_____

Next he uses them all to make the smallest 4-digit number he can.

c) Write it in digits. _____

d) Write it in words. _____

_____

4 marks

2 | How much would you have if you had one of each type of UK coin?

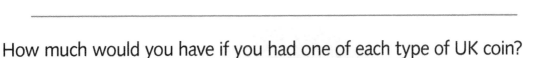

Write the amount in words. _____

1 mark

Marks.......... /5

Total marks .... /19

How am I doing?

10 and 100 More or Less

PS Problem-solving questions

Challenge 1

1 Increase these prices by 10p.

a)  68p

————p

b)  19p

————p

c)  73p

————p

3 marks

2 Decrease these prices by 10p.

a)  27p

————p

b)  51p

————p

c) 44p

————p

3 marks

3 Sara has torn these scraps from her 100 square. Fill in the missing numbers.

a)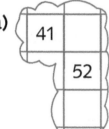

| 41 | |
| | 52 |

b)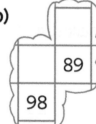

| | 89 |
| 98 | |

c)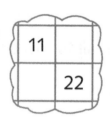

| 11 | |
| | 22 |

d)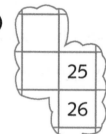

| | 25 |
| | 26 |

4 marks

Marks.........../10

18

10 and 100 More or Less

Challenge 2

1 Complete the gaps in this table.

10 less	Start number	10 more
60	70	
	30	40
	25	
45		65
5		

7 marks

2 Find the missing weights of these animals.

a) –100 kg   +100 kg

[ ] | 695 kg | [ ]

b)  –100 kg    +100 kg

[ ] | 310 kg | [ ]

4 marks

Marks.......... /11

Challenge 3

PS **1** Look at this number box.

IN ⟶ | –10 ⟶ +100 ⟶ –10 ⟶ +100 | ⟶ OUT

a) If 105 goes into the box, what number comes out? _____

b) If 131 goes into the box, what number comes out? _____

c) 203 came **out** of the box. What number went in? _____

3 marks

Marks.......... /3

Total marks .... /24   How am I doing?

Counting in Multiples

Challenge 1

1 Use multiples of 4, 8, 50 and 100 to complete the missing stages of these rockets before blast off.

a)
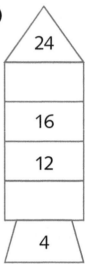
| 24 |
| 16 |
| 12 |
| |
| 4 |

b)
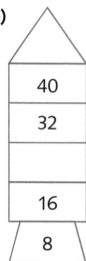
| |
| 40 |
| 32 |
| |
| 16 |
| 8 |

c)
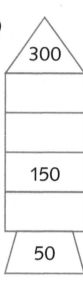
| 300 |
| |
| 150 |
| |
| 50 |

d)

| |
| 500 |
| |
| 200 |
| 100 |

4 marks

Marks.......... /4

Challenge 2

1 Look at these rows of flags. Which multiples do they show?

a)
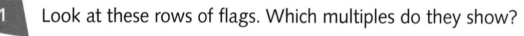
56 64 72 80   Multiples of ☐

b)

400 350 300 250   Multiples of ☐

c)

32 36 40 44   Multiples of ☐

d)

24 20 16 12   Multiples of ☐

4 marks

Counting in Multiples

2 Write the next three multiples. Be careful: sometimes the numbers are getting smaller.

a) 800, 700, 600, _____, _____, _____

b) 16, 20, 24, _____, _____, _____

c) 96, 88, 80, _____, _____, _____

d) 850, 900, 950, _____, _____, _____

e) 88, 92, 96, _____, _____, _____

f) 16, 24, 32, _____, _____, _____

6 marks

Marks.........../10

Challenge 3

1 Colour two routes across this hexagon grid. One goes up in multiples of 4 and the other goes up in multiples of 8.

Multiples of 4 start here

Multiples of 8 start here

68   64   56

62   72

76   58

78   80   64

88   72

96   84   70

90   88

104   82   92

Multiples of 8 end here

Multiples of 4 end here

2 marks

2 a) Write a number that is a multiple of 4 and 50. _____

b) Write a number that is a multiple of 8 and 100. _____

2 marks

Marks.........../4

Total marks ..../18

How am I doing?

21

Comparing and Ordering Numbers

 **PS** Problem-solving questions

Challenge 1

1 Put these numbers in order from lowest to highest.

a) 106   161   116   160   166

_____ _____ _____ _____ _____

1 mark

2 Write a number that could come between these pairs of numbers.

a) 198   202

b) 245   254

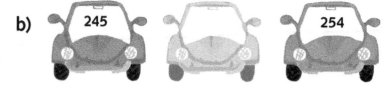

2 marks

3 Fill in the missing numbers.

a)
| 96 | 97 | 98 | | | | 102 | 103 | 104 | |

b)
| 383 | 384 | | | 387 | | 389 | | | |

2 marks

Marks.......... /5

Challenge 2

1 Label these snakes 1–5 to put them in order from shortest to longest.

118 cm   188 cm   108 cm   180 cm   158 cm

☐ ☐ ☐ ☐ ☐

1 mark

Comparing and Ordering Numbers

2 Label these supermarket items 1–5 to put them in order from cheapest to most expensive.

£1.05   £2.29   £1.99   £1.12   £2.15

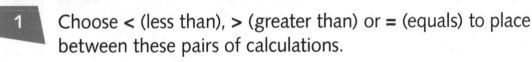

☐   ☐   ☐   ☐   ☐

1 mark

3 Choose either the 'greater than' **>** or the 'less than' **<** symbol to place between these pairs of numbers or amounts.

a) 56 ☐ 65   **b)** 432 ☐ 423

c) 303 cm ☐ 390 cm   **d)** £6.77 ☐ £6.66

4 marks

Marks.......... /6

Challenge 3

1 Choose **<** (less than), **>** (greater than) or **=** (equals) to place between these pairs of calculations.

a) 16 – 4 ☐ 8 + 4   **b)** 4 × 8 ☐ 7 × 5

c) 5 + 6 ☐ 20 ÷ 2   **d)** 108 – 50 ☐ 6 × 8

4 marks

PS **2** Scarlett has these three digit cards. ⑤ ② ⑨

She uses them to make six different 3-digit numbers, which she writes in order from smallest to largest.

Write Scarlett's numbers here.

smallest _____ _____ _____ _____ _____ **largest**

7 marks

Marks.......... /11

Total marks .... /22   How am I doing?

Number Problems

Challenge 1

PS 1 Class 3 are growing sunflowers. Lola's sunflower is 35 cm tall. Caamir's is 10 cm taller. Blanka's is 10 cm shorter. Write the heights of their sunflowers.

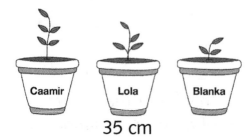

⬜ cm   35 cm   ⬜ cm

2 marks

PS 2 Luca ran 150 m. Joe ran 100 m less and Liu ran 100 m more. How far did Joe and Liu run?

Joe _____ m   Liu _____ m

2 marks

PS 3 **a)** In Class 3, the children sit in groups of 4. There are 24 children in the class. How many groups is this? _____

b) The school secretary buys envelopes in boxes of 100. She has 6 boxes in her cupboard. How many envelopes is this? _____

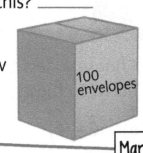

100 envelopes

2 marks

Marks.......... /6

Challenge 2

PS 1 **a)** George is 10 cm shorter than Anna. Anna is 108 cm tall.

How tall is George? _____ cm

b) Karol is 10 cm taller than Anna. How tall is Karol? _____ cm

2 marks

PS 2 **a)** Jaydip has 100 more football cards than Lucy. Lucy has 161 cards.

How many cards does Jaydip have? _____

b) Rayhan has 100 fewer cards than Lucy.

How many cards does Rayhan have? _____

2 marks

Number Problems

PS **3**

a) Pencils come in boxes of 50. Miss Patel has 5 boxes in her cupboard.

How many pencils is this? _____

50 pencils

b) Scissors come in boxes of 8. Miss Patel needs 32 pairs for her class.

How many boxes should she order? _____

2 marks

Marks.......... /6

Challenge 3

PS **1**

a) Amelie was born in 2009. Her brother is 10 years older.

When was he born? _____

b) Amelie's great grandfather was born 100 years before Amelie.

When was he born? _____

2 marks

2 James has four coins in his pocket. They are a mixture of 50p and £1 coins.

a) Circle the amount of money that James could NOT have.

£3.50   £1.50   £3.00   £2.50

b) James says that he has less than £4 but more than £3.

How much money does James have? _____

2 marks

Marks.......... /4

Total marks .... /16   How am I doing?

Adding and Subtracting Mentally

Challenge 1

1 Do these additions mentally. Write the answers.

a) $43 + 6 = \boxed{}$ b) $15 + 6 = \boxed{}$ c) $122 + 9 = \boxed{}$

d) $235 + 5 = \boxed{}$ e) $198 + 6 = \boxed{}$ f) $817 + 8 = \boxed{}$

6 marks

2 Do these subtractions mentally. Write the answers.

a) $18 - 4 = \boxed{}$ b) $76 - 7 = \boxed{}$ c) $345 - 6 = \boxed{}$

d) $103 - 9 = \boxed{}$ e) $555 - 7 = \boxed{}$ f) $761 - 8 = \boxed{}$

6 marks

Marks.........../12

Challenge 2

1 Do these additions mentally. Write the answers.

a) $135 + 10 = \boxed{}$ b) $293 + 25 = \boxed{}$

c) $612 + 219 = \boxed{}$

3 marks

2 Do these subtractions mentally. Write the answers.

a) $250 - 70 = \boxed{}$ b) $182 - 7 = \boxed{}$ c) $508 - 15 = \boxed{}$

3 marks

Marks.........../6

Challenge 3

1 Work out these additions by partitioning the numbers into tens and ones before adding.

Example: $423 + 138 = (400 + 100) + (20 + 30) + (3 + 8)$
$= 500 + 50 + 11 = 561$

Adding and Subtracting Mentally

a) 267 + 252 = (_____ + _____) + (_____ + _____)

+ (_____ + _____) = _____ + _____ + _____ = _____

b) 514 + 388 = (_____ + _____) + (_____ + _____)

+ (_____ + _____) = _____ + _____ + _____ = _____

2 marks

2 Work out these subtractions by counting up from the smaller to the larger number to find the difference.

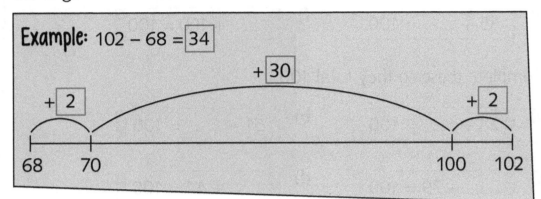

Example: 102 – 68 = 34

a) 507 – 428 = ☐

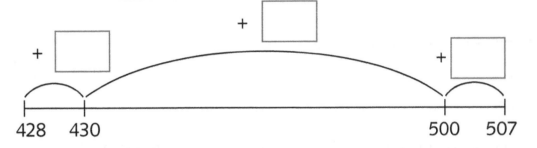

b) 809 – 742 = ☐

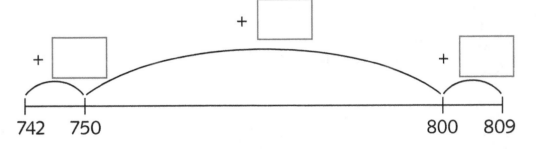

2 marks

Marks.......... /4

Total marks .... /22

How am I doing?

Number Bonds

Challenge 1

1 Complete these so they total 100.

a) 20 + _____ = 100

b) _____ + 75 = 100

c) _____ + 5 = 100

d) 55 + _____ = 100

e) 35 + _____ = 100

f) _____ + 100 = 100

6 marks

2 Complete these so they total 100.

a) 28 + _____ = 100

b) 51 + _____ = 100

c) _____ + 79 = 100

d) _____ + 44 = 100

e) 7 + _____ = 100

f) 32 + _____ = 100

6 marks

Marks........../12

Challenge 2

PS **1** George is going shopping. What would his change from £1 be if he bought these items?

a)  69p _____p

b)  22p _____p

c) 73p _____p

d) 81p _____p

4 marks

Number Bonds

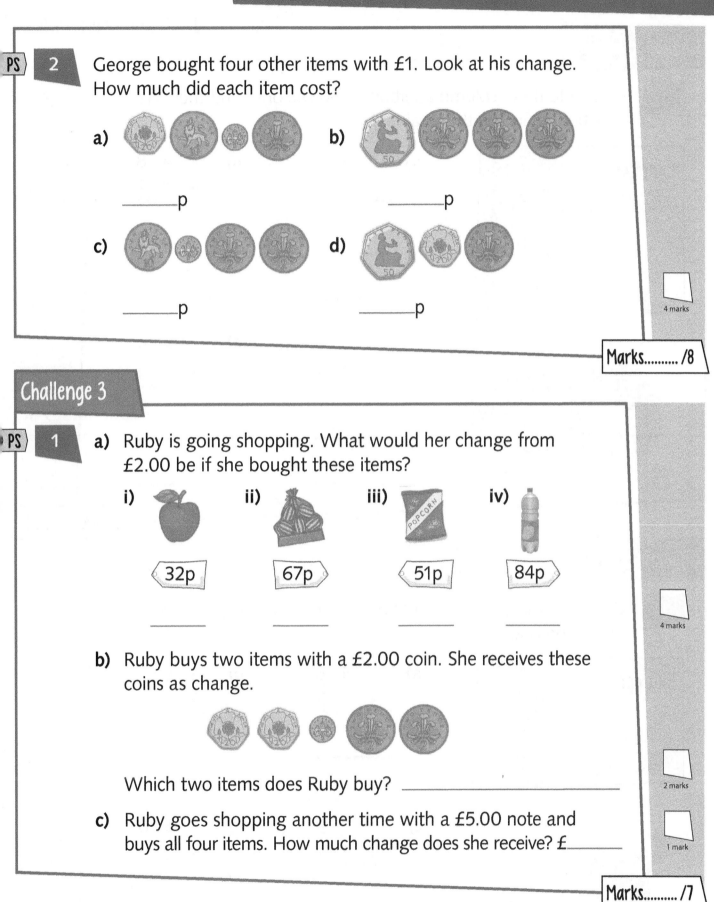

PS **2** George bought four other items with £1. Look at his change. How much did each item cost?

a) _____p

b) _____p

c) _____p

d) _____p

4 marks

Marks.......... /8

Challenge 3

PS **1** a) Ruby is going shopping. What would her change from £2.00 be if she bought these items?

i) 32p

ii) 67p

iii) 51p

iv) 84p

_____ _____ _____ _____

4 marks

b) Ruby buys two items with a £2.00 coin. She receives these coins as change.

Which two items does Ruby buy? _____

2 marks

c) Ruby goes shopping another time with a £5.00 note and buys all four items. How much change does she receive? £_____

1 mark

Marks.......... /7

Total marks .... /27

How am I doing?

Adding in Columns

PS Problem-solving questions

Challenge 1

1 Complete these column additions. Add the ones and then the tens before finding the total.

Example:
```
      2 3
  +   3 4
  ─────────
        7
      5 0
  ─────────
      5 7
```

a)
```
      6 2
  +   1 7
  ─────────


  ─────────
```

b)
```
      4 8
  +   4 2
  ─────────


  ─────────
```

c)
```
      1 5
  +   7 7
  ─────────


  ─────────
```

d)
```
      5 6
  +   3 3
  ─────────


  ─────────
```

4 marks

Marks.......... /4

Challenge 2

1 Complete these column additions. Carry any tens you make below the line.

Example:
```
      1 7
  +   8 5
  ─────────
    1 0 2
        1
```

a)
```
      4 6
  +   4 2
  ─────────

```

b)
```
      3 6
      2 7
  ─────────

```

c)
```
      7 9
  +   1 1
  ─────────

```

d)
```
      5 0
  +   2 9
  ─────────

```

4 marks

Adding in Columns

2 Now complete these column additions with hundreds as well as tens and ones.

Example:

```
    4 6 4
  + 2 5 5
  -------
    7 1 9
  -------
      1
```

a)
```
    1 3 2
  + 1 5 6
  -------
```

b)
```
    5 1 8
  +   4 9
  -------
```

c)
```
    6 6 2
  + 2 8 6
  -------
```

d)
```
      3 6
  + 7 4 6
  -------
```

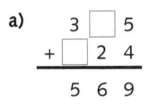

4 marks

Marks.......... /8

Challenge 3

PS **1** Write the missing digits in these column additions.

a)
```
    3 □ 5
  + □ 2 4
  -------
    5 6 9
  -------
```

b)
```
    □ 0 □
  + 2 □ 8
  -------
    6 4 0
  -------
```

c)
```
      6 □
  + □ □ 7
  -------
    1 1 5
  -------
```

d)
```
    □ □ 0
  + 1 7 □
  -------
    6 3 0
  -------
```

e)
```
    2 □ 5
  + □ 1 □
  -------
    9 3 4
  -------
```

5 marks

Marks.......... /5

Total marks .... /17

How am I doing?

Subtracting in Columns

PS Problem-solving questions

Challenge 1

1 Complete these column subtractions.

a)
```
    7 8
  - 3 4
  _____
```

b)
```
    8 5
  - 4 2
  _____
```

c)
```
  9 3 8
  -   1 4
  _____
```

d)
```
  3 7 9
  - 2 5 4
  _____
```

e)
```
  3 5 9
  -   3 8
  _____
```

f)
```
  7 6 7
  - 4 0 5
  _____
```

g)
```
  5 9 1
  - 2 6 1
  _____
```

h)
```
  4 7 4
  -   3 1
  _____
```

8 marks

Marks.......... /8

Challenge 2

1 Complete these column subtractions. Be careful: you may have to exchange a ten.

Example:
```
      4 1
  6 5̸ 0
  - 2 1 7
  _____
    4 3 3
```

a)
```
  8 9 6
  - 5 7 5
  _____
```

b)
```
  1 8 8
  -   6 9
  _____
```

c)
```
  3 7 2
  - 1 2 8
  _____
```

d)
```
  8 9 3
  -   7 5
  _____
```

4 marks

Subtracting in Columns

PS **2** Write the missing digits in these column subtractions.

a)
```
    □ 6
 -  3 □
 _____
    2 1
```

b)
```
    8 □
 -  □ 2
 _____
    7 7
```

c)
```
    □ 5 0
 -  □ 0
 _____
    3 2 0
```

d)
```
    9 □ 4
 -  □ 3 □
 _____
    2 1 2
```

e)
```
    2 6 □
 -  □ 4 5
 _____
    1 1 8
```

5 marks

Marks......... /9

Challenge 3

1 Complete these column subtractions. Be careful: you may have to exchange a ten or a hundred.

Example:
```
    ⁴5̸ ¹2 3
 -   1 8 1
 _____
     3 4 2
```

a)
```
    6 0 9
 -  1 2 7
 _____
```

b)
```
    5 5 5
 -  1 9 4
 _____
```

c)
```
    8 2 3
 -    5 7
 _____
```

d)
```
    4 4 6
 -  2 8 8
 _____
```

4 marks

Marks......... /4

Total marks .... /21   How am I doing?

Estimating and Checking Calculations

Challenge 1

1 Draw a circle around the number or measurement that is the best estimate.

a) Pages in a magazine

   6   60   600

b) Beans in a tin

   50   5   500

c) Your height

   120 cm   12 cm   1200 cm

d) Popcorn in a packet

   70   700   7000

e) Weight of a baby

   50 kg   500 kg   5 kg

5 marks

2 Circle the inverse calculation you could use to check if each calculation is correct.

a) | 25 – 6 = 19 |

   19 + 6 = 25   5 × 5 = 25   10 + 9 = 19

b) | 65 + 65 = 130 |

   65 – 5 = 60   130 – 65 = 65   130 ÷ 65 = 2

c) | £1.00 – 55p = 45p |

   £1.00 + 55p = £1.55   £1.00 + 45p = £1.45   45p + 55p = £1.00

3 marks

Marks.......... /8

Estimating and Checking Calculations

Challenge 2

1 Circle the number you estimate to be closest to the real answer.

a) 31×5 150   15   1500

b) $139 + 156$ 30   3000   300

c) $1000 - 508$ 5000   500   50

d) $100 \div 23$ 4   400   40

4 marks

2 Write the inverse calculation you could use to check if each calculation is correct.

a) $35 - 24 = 11$ b) $14 \times 10 = 140$

_____   _____

c) $120 \div 4 = 30$ d) $570 + 310 = 880$

_____   _____

4 marks

Marks.......... /8

Challenge 3

1 The label on each jar shows how many sweets each one holds when full. Estimate how many sweets are left in each jar.

a) _____    b) _____    c) _____

30   160   180

3 marks

Marks.......... /3

Total marks .... /19   How am I doing?

Addition and Subtraction Problems

Challenge 1

PS **1** Alison, Jane and Johnny are selling cakes at the school fair. Alison sells 125 cakes. Jane sells 15 more and Johnny sells 30 fewer.

How many cakes do they sell?

a) Jane ☐ **b)** Johnny ☐ **c)** Altogether ☐

3 marks

PS **2** At Treetops Primary School there are 210 children. At Ash Grove Primary there are 360 children.

a) How many more children are there at Ash Grove than at Treetops? ☐

b) How many children are there altogether? ☐

2 marks

Marks.......... /5

Challenge 2

PS **1** Perry is 147 cm tall. Kitty is 13 cm shorter and Harry is 8 cm taller.

a) How tall is Kitty? _____ cm

b) How tall is Harry? _____ cm

c) How much taller is Harry than Kitty? _____ cm

3 marks

PS **2** This table shows how many children attend lunchtime clubs.

Choir	Coding	Chess	Netball	Book club
23	8	12	17	9

a) How many children attend the clubs altogether? _____

b) How many more children attend choir than coding? _____

2 marks

PS **3** Benji spends £100 on football boots, shorts, socks and a shirt. Boots cost £38, shorts £18 and socks £8.

How much does the shirt cost? £ _____

1 mark

36

Addition and Subtraction Problems

PS **4** In a 'magic square' every row and every column totals 15 using the numbers 1–9 only once each. Fill in the missing numbers in these magic squares.

a)

	1	8
7		3

b)

		2
1	5	
	3	

c)

8		
	5	
	9	2

3 marks

Marks.......... /9

Challenge 3

PS **1** **a)** Jake buys a drink that costs 78p. He pays with a pound and receives three coins as change. What could his coins be?

_____ + _____ + _____

b) Leah buys a packet of football cards. She pays with a £2 coin and receives the change shown here.

How much were the cards? _____

2 marks

PS **2** In an addition pyramid, each number is the total of the two numbers below it. Complete the missing blocks in this pyramid.

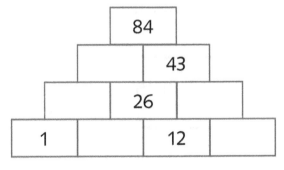

5 marks

Marks.......... /7

Total marks .... /21

How am I doing?

1. Write the entry numbers on these cyclists' helmets from the smallest to largest.

505   498   489   551   515

1 mark

2. Write the missing number on each pair to make 100.

a)   b)   c)   d)

47   13   59   74

4 marks

3. Complete these column additions.

a)
```
    4 7
+ 9 3 5
───────

───────
```

b)
```
  7 1 9
+ 1 1 8
───────

───────
```

c)
```
  3 0 7
+ 5 9 2
───────

───────
```

d)
```
  6 4 5
+   7 3
───────

───────
```

4 marks

PS **4.** A snail crawls 198 cm. A tortoise crawls 205 cm. How much further did the tortoise crawl than the snail? _____

1 mark

5. Complete these addition grids as quickly as you can.

a)

+	75	
40		125
55		

b)

+		65
80		
	160	130

2 marks

6. Complete these column subtractions.

a)

```
    5 8
  − 4 1
  _____

  _____
```

b)

```
    9 4
  − 7 2
  _____

  _____
```

c)

```
    4 2
  − 2 3
  _____

  _____
```

d)

```
    2 0 5
  −   5 3
  _____

  _____
```

4 marks

7. Partition these numbers into hundreds, tens and ones.

a) 672 = ☐ + ☐ + ☐

b) 123 = ☐ + ☐ + ☐

2 marks

8. What numbers are the arrows pointing to?

a) [   ]   b) [   ]   c) [   ]

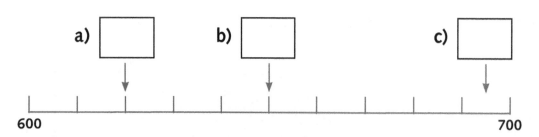

600   700

3 marks

PS ⟩ **9.** Izzy has only 10p and 1p coins in her purse. She has four coins. Circle the amount of money that Izzy could NOT have.

31p   13p   32p   22p

1 mark

10. Write these numbers in digits.

a) Ninety-seven   [   ]

b) One hundred and four   [   ]

c) Three hundred and fifteen [   ]

d) Three hundred and fifty-five [   ]

4 marks

11. Write the next three multiples.

a) 28, 32, 36, _____, _____, _____

b) 80, 72, 64, _____, _____, _____

2 marks

PS ⟩ **12.** Archie, Gus and Beth are picking strawberries. Archie picks 205 strawberries. Gus picks 30 more and Beth picks 6 fewer.

a) How many strawberries does Beth pick? _____

b) How many strawberries do they pick altogether? _____

2 marks

13. Neil buys a kilogram of each fruit. Circle the best estimate of how many of each fruit he has bought.

a)  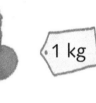 **5, 50** or **500** strawberries

b) **8, 80** or **800** bananas

c) **20, 200** or **2000** cherries

3 marks

14. At ScreenSave a tablet computer costs £295.

£295

How much does it cost at these stores?

a) At ByteSize it is £10 more. _____

b) At DigiMarket it is £100 less. _____

2 marks

Marks........ /35

Multiplication and Division Facts

PS Problem-solving questions

Challenge 1

1 Write two multiplication and two division facts for each array.

a)

_____ × _____ = _____

_____ × _____ = _____

_____ ÷ _____ = _____

_____ ÷ _____ = _____

b)

_____ × _____ = _____

_____ × _____ = _____

_____ ÷ _____ = _____

_____ ÷ _____ = _____

4 marks

2 Complete these multiplication grids.

a)

×	3	10	2
5			
2			
4			

b)

×	3	4	5
2			
8			
3			

2 marks

Marks.......... /6

Challenge 2

1 Write two multiplication and two division facts for each set of three numbers.

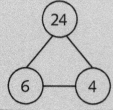

Example:

$6 \times \underline{4} = \underline{24}$ $\underline{24} \div \underline{4} = \underline{6}$

$\underline{4} \times \underline{6} = \underline{24}$ $\underline{24} \div \underline{6} = \underline{4}$

Multiplication and Division Facts

a)

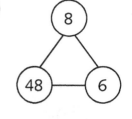

____ × ____ = ____

____ × ____ = ____

____ ÷ ____ = ____

____ ÷ ____ = ____

b)

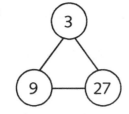

____ × ____ = ____

____ × ____ = ____

____ ÷ ____ = ____

____ ÷ ____ = ____

c)

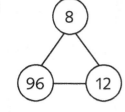

____ × ____ = ____

____ × ____ = ____

____ ÷ ____ = ____

____ ÷ ____ = ____

6 marks

Marks.......... /6

Challenge 3

PS **1** Complete these multiplication grids.

a)

×	8		4
2			
	64		
5		15	

b)

×			
		16	
10	40	20	30
4			

2 marks

2 Use the multiplication facts you know to solve these.

a) 320 ÷ 4 = _____

b) 60 × 5 = _____

c) 8 × 80 = _____

d) 220 ÷ 11 = _____

e) 960 ÷ _____ = 12

f) 40 × _____ = 1600

g) _____ ÷ 8 = 50

h) 30 × 30 = _____

8 marks

Marks......... /10

Total marks ..../22   How am I doing?

Doubling and Halving

Challenge 1

1 Draw lines to join each number to its double.

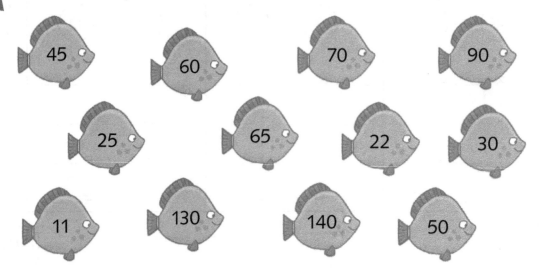

6 marks

2 Write in the numbers to complete these sentences.

a) Half of eighty is _____. b) Half of _____ is thirty.

c) Double _____ is seventy. d) Half of forty-four is _____.

e) Double eighty is _____. f) Double twelve is _____.

6 marks

Marks.........../12

Challenge 2

PS **1** This machine doubles any number put into it. Complete the missing inputs and outputs.

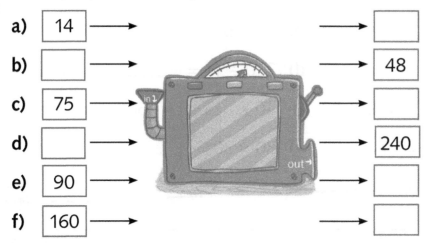

a) 14 ⟶   ⟶ ☐

b) ☐ ⟶   ⟶ 48

c) 75 ⟶   ⟶ ☐

d) ☐ ⟶   ⟶ 240

e) 90 ⟶   ⟶ ☐

f) 160 ⟶   ⟶ ☐

6 marks

Doubling and Halving

PS **2** This is Bilal's pocket money.

His older sister, Akila, gets double this and his younger brother, Zain, gets half.

How much pocket money do they get?

a) Bilal _____ **b)** Zain _____ **c)** Akila _____

3 marks

Marks.......... /9

Challenge 3

1 Use doubling to show the link between the 2×, 4× and 8× tables.

a) $8 =$ _____ $\times 8 =$ _____ $\times 4 =$ _____ $\times 2$

b) $16 =$ _____ $\times 8 =$ _____ $\times 4 =$ _____ $\times 2$

c) $24 =$ _____ $\times 8 =$ _____ $\times 4 =$ _____ $\times 2$

3 marks

PS **2** Xavi swims 200 metres. He then cycles twice this distance. Finally he runs half the distance he has swum and cycled. How far does Xavi travel altogether?

_____

1 mark

PS **3** Miss Peake spent half of her money on a sandwich. Half of what she had left she spent on a drink, which cost 50p. How much money did Miss Peake start with?

_____

1 mark

Marks.......... /5

Total marks .... /26   How am I doing?

3×, 4× and 8× Tables

Challenge 1

1 Fill in the gaps in this multiplication table chart.

	×1	×2	×3	×4	×5	×6	×7	×8	×9	×10	×11	×12
3												
4												
8												

3 marks

2 Now use your chart to answer these questions.

a) $2 \times 8 =$ _____   b) $10 \times 4 =$ _____

c) $6 \times 3 =$ _____   d) $3 \times 3 =$ _____

e) $12 \times 2 =$ _____   f) $7 \times 8 =$ _____

6 marks

PS **3** Hanita buys four bunches of flowers.
Each bunch has eight flowers.

How many flowers has she bought? _____

1 mark

PS **4** In class 3P there are seven tables of four children.

How many children are in the class? _____

1 mark

Marks.......... /11

Challenge 2

1 Complete these using your times tables.

a) $5 \times$ _____ $= 15$ b) _____ $\times 3 = 12$

c) $12 \times 8 =$ _____   d) _____ $\times 4 = 28$

e) $8 \times 8 =$ _____   f) $11 \times$ _____ $= 44$

6 marks

3×, 4× and 8× Tables

PS **2** Mr Clarke is ordering sports equipment for the school.

£4   £8   £3

a) How much would six tennis racquets cost? _____

b) How much would nine cricket bats cost? _____

c) Which costs more, nine footballs or five tennis racquets? _____

d) How many tennis racquets can he buy with £80? _____

4 marks

Marks......... /10

Challenge 3

PS **1** Put these numbers into the correct zone in this Venn diagram.

| 8 | 9 | 12 | 13 | 15 | 20 | 24 |

Numbers in the 3× table   Numbers in the 4× table

2 marks

Marks.......... /2

Total marks .... /23   How am I doing?

Mental Multiplication and Division

Challenge 1

1 Solve these multiplications. Then use coloured pencils to colour pairs of bricks with the same answer.

$8 \times 5 \times 2 =$ _____	$10 \times 3 \times 2 =$ _____	$4 \times 3 \times 2 =$ _____

$5 \times 2 \times 4 =$ _____	$5 \times 8 \times 2 =$ _____	

$2 \times 3 \times 10 =$ _____	$2 \times 4 \times 3 =$ _____	$4 \times 5 \times 2 =$ _____

8 marks

2 Place <, > or = between these calculations to make each number statement correct.

a) $10 \div 2$ ☐ $12 \div 4$

b) $16 \div 4$ ☐ $32 \div 8$

c) 7×3 ☐ 6×4

d) $18 \div 6$ ☐ $24 \div 8$

4 marks

Marks.........../12

Challenge 2

1 Use each calculation to solve the two related ones behind.

a)

$5 \times 3 = 15$ $5 \times 30 =$ _____ $50 \times 3 =$ _____

b)
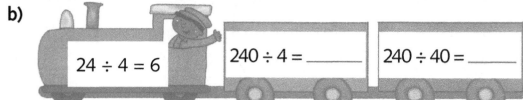
$24 \div 4 = 6$ $240 \div 4 =$ _____ $240 \div 40 =$ _____

c)

$4 = 8 \div 2$ _____ $= 80 \div 2$ _____ $= 80 \div 20$

Mental Multiplication and Division

d)

$40 \times 3 =$ _____ $4 \times 30 =$ _____

4 marks

2 Place **<**, **>** or **=** between these calculations to make each number statement correct.

a) $48 \div 8$ ☐ $21 \div 3$ **b)** $240 \div 2$ ☐ 5×30

c) 40×4 ☐ 90×2 **d)** 11×30 ☐ 3×110

4 marks

Marks.......... /8

Challenge 3

1 Switch numbers around in these multiplications to make them easier to solve. Part of the first one has been done for you.

a) $4 \times 12 \times 2 =$ __4__ $\times$ __2__ $\times 12 =$ __8__ $\times 12 =$ _____

b) $2 \times 16 \times 5 =$ _____ $\times$ _____ $\times 16 =$ _____ $\times 16 =$ _____

c) $3 \times 9 \times 2 =$ _____ $\times$ _____ $\times 9 \ =$ _____ $\times 9 \ =$ _____

3 marks

2 Find two related calculations for each of these multiplications and divisions.

a) $8 \times 4 = 32$ ____ $\times$ ____ $=$ ____   ____ $\times$ ____ $=$ ____

b) $36 \div 12 = 3$ ____ $\div$ ____ $=$ ____   ____ $\div$ ____ $=$ ____

c) $7 \times 4 = 28$ ____ $\times$ ____ $=$ ____   ____ $\times$ ____ $=$ ____

3 marks

Marks.......... /6

Total marks .... /26   How am I doing?

Multiplying 2-digit Numbers

PS Problem-solving questions

Challenge 1

1 Multiply these numbers by 10.

a) $23 \times 10 =$ _____   b) $12 \times 10 =$ _____   c) $51 \times 10 =$ _____

d) $88 \times 10 =$ _____   e) $49 \times 10 =$ _____   f) $17 \times 10 =$ _____

6 marks

2 Partition these numbers before multiplying.

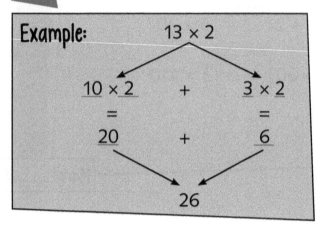

Example:

$$13 \times 2$$

$$10 \times 2 \quad + \quad 3 \times 2$$
$$= \qquad =$$
$$20 \quad + \quad 6$$
$$26$$

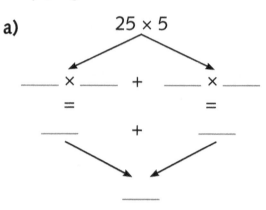

a) 25×5

_____ $\times$ _____ $+$ _____ $\times$ _____
$= \qquad =$
_____ $+$ _____
_____

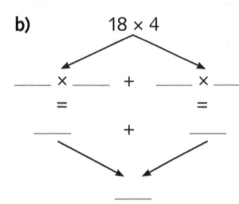

b) 18×4

_____ $\times$ _____ $+$ _____ $\times$ _____
$= \qquad =$
_____ $+$ _____
_____

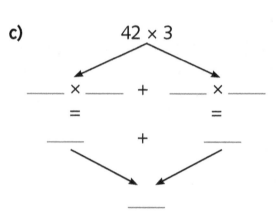

c) 42×3

_____ $\times$ _____ $+$ _____ $\times$ _____
$= \qquad =$
_____ $+$ _____
_____

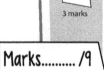
3 marks

Marks.......... /9

Challenge 2

1 Use these grids to help you multiply these 2-digit numbers.

Example: $64 \times 4 = \underline{256}$

×	60	4
4	240	16

Multiplying 2-digit Numbers

a) $27 \times 8 =$ _____

×		

b) $32 \times 6 =$ _____

×		

2 marks

PS **2** Pencils come in boxes of 24. Mr Knight has 5 boxes. How many pencils does he have? _____

1 mark

PS **3** Three buses carry 58 children and adults each. How many is this altogether? _____

1 mark

Marks.......... /4

Challenge 3

1 Use these grids to help you multiply these 3-digit numbers.

a) $142 \times 5 =$ _____

×			

b) $231 \times 4 =$ _____

×			

2 marks

PS **2** This half-term, Mumtaz spent 20p a day on fruit for thirty-five days. How much did she spend? _____

1 mark

PS **3** At the school office, Mrs Ward collects 25p a week from each child for healthy snacks. How much will she have collected from one child after six weeks? _____

1 mark

Marks.......... /4

Total marks .... /17   How am I doing?

Multiplication and Division Problems

PS Problem-solving questions

Challenge 1

PS **1** There are 10 football cards in a pack.

  a) How many cards will there be in 3 packs? _____

  b) How many cards will there be in 6 packs? _____

  c) Dan has 50 cards. How many packs has he bought? _____

3 marks

PS **2** Look at this picture of Dance and Music club.

How many groups will there be if they have to perform in these groupings?

  a) Sixes _____   **b)** Threes _____   **c)** Fours _____

3 marks

Marks......... /6

Challenge 2

PS **1** Complete these problems. The first one has been done for you.

Problem	Number statement
18 children go on a canoe trip. One canoe takes 3 children. How many canoes do they need?	$18 \div 3 = 6$
a) Afrah is 4 times older than her daughter, Kalima. Kalima is 8. How old is Afrah?	_____
b) James spent £1.20 at the cake shop on tea cakes costing 40p. How many did he buy?	_____
c) Logan gets twice as much pocket money as Patrick. Patrick gets 80p. How much does Logan get?	_____

3 marks

Multiplication and Division Problems

PS **2** Jake has 38p to spend on biscuits that cost 8p each.

How many biscuits can Jake afford? _____

1 mark

PS **3** At the hostel, 4 children sleep in each room. There are 30 children in the class.

How many rooms do they need? _____

1 mark

Marks.......... /5

Challenge 3

PS **1** During school assembly there are 30 children performing on the stage. There are twice as many children as that sitting on benches. There are 4 times as many children sitting on the floor as on the stage.

a) How many children are sitting on benches? _____

b) How many children are sitting on the floor? _____

c) How many children are there in the assembly altogether? _____

d) For every 30 children in the hall there is one adult. How many adults are there? _____

4 marks

PS **2** Every evening Swim School has up to 20 pupils.

a) On Monday, the teacher put the children into 3s and there were 2 left over. When she put the children into 4s there were 3 left over. How many children were there?

b) On Tuesday, there were 2 children left over when she put the children into 3s **and** into 4s. How many children were there?

2 marks

Marks.......... /6

Total marks .... /17   How am I doing?   😊   😐   ☹

Tenths

Challenge 1

1 Colour the shapes to show these fractions.

a) $\frac{5}{10}$

b) $\frac{3}{10}$

c) $\frac{6}{10}$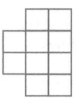

 3 marks

 2 Lewis has bought a bag of 10 peppers.

$\frac{4}{10}$ are green. $\frac{1}{10}$ are orange. $\frac{2}{10}$ are yellow. The rest are red.

a) Colour the peppers correctly.

b) What fraction are red?

2 marks

Marks.......... /5

Challenge 2

1 Fill in the missing fractions on this number line.

| 0 | $\frac{1}{10}$ | $\frac{2}{10}$ | | $\frac{4}{10}$ | $\frac{5}{10}$ | | | $\frac{8}{10}$ | | 1 |

4 marks

2 Divide these quantities by 10 to find a tenth.

a) $\frac{1}{10}$ of 20 = _____   b) $\frac{1}{10}$ of 70 = _____

c) $\frac{1}{10}$ of £1.00 = _____   d) $\frac{1}{10}$ of 30 cm = _____

e) $\frac{1}{10}$ of 80 kg = _____   f) $\frac{1}{10}$ of 100 km = _____

6 marks

Marks........ /10

Tenths

Challenge 3

PS **1** Cameron has read $\frac{4}{10}$ of his book, which has 800 pages. Nancy has read $\frac{7}{10}$ of her book, which has 400 pages. How many more pages has Cameron read than Nancy?

_____

1 mark

2 Complete these by dividing by 10. Be careful: the question is not always asking for **one** tenth.

a) $\frac{1}{10}$ of £400 = _____

b) $\frac{1}{10}$ of 650 m = _____

c) $\frac{4}{10}$ of 20 = _____

d) $\frac{1}{10}$ of 50 = _____

e) $\frac{1}{10}$ of 330 = _____

f) $\frac{7}{10}$ of 100 = _____

6 marks

PS **3** In Class 3 there are 30 children.

a) $\frac{6}{10}$ are boys. What number of children in the class are girls?

_____

b) $\frac{1}{10}$ wear glasses. What number of children do not wear glasses? _____

c) 9 children have packed lunches. How many tenths is this?

_____

3 marks

Marks.........../10

Total marks ..../25   How am I doing?

Recognising Fractions

Challenge 1

1 What fraction of each shape is shaded?

a)

b)

2 marks

2 What fractions are the arrows pointing to on this number line?

a) ☐   b) ☐   c) ☐

0   1

3 marks

PS **3** Sian has a bead necklace. Half of the beads are yellow, one quarter are red and one quarter are blue. Colour Sian's necklace.

3 marks

Marks.......... /8

Challenge 2

1 What fraction of each shape is shaded?

a)

b)

2 marks

2 What fractions are the arrows pointing to on this number line?

a) ☐   b) ☐   c) ☐

0   1

3 marks

Recognising Fractions

3 In a traffic jam there are cars and lorries.

a) What fraction of the vehicles are cars? ☐

b) What fraction of the vehicles are lorries? ☐

2 marks

Marks.........../7

Challenge 3

PS **1** Complete the shading so that:

a) $\frac{1}{2}$ is shaded.

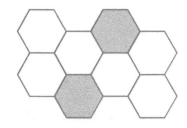

b) $\frac{3}{4}$ is shaded.

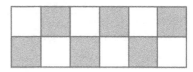

2 marks

2 What fractions or mixed numbers are the arrows pointing to on this number line?

a) ☐ **b)** ☐ **c)** ☐

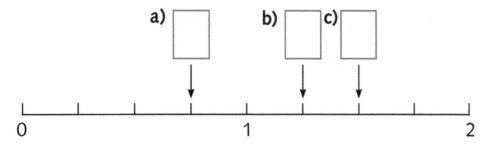

3 marks

PS **3** Karl has completed 20 km of an 80 km journey.

What fraction is this? ☐

1 mark

Marks.........../6

Total marks ..../21   How am I doing? 😊 😐 ☹

57

Fractions of Amounts

Challenge 1

1 Look at these planes.

a) Draw a line around half of them and colour them orange.

b) Draw a line around a quarter of them and colour them blue.

c) Complete this sentence about the planes.

$\frac{1}{2}$ of _____ is _____ and $\frac{1}{4}$ of _____ is _____.

3 marks

2 Look at these apples.

a) Draw a line around a quarter of them and colour them red.

b) Draw a line around three quarters of them and colour them green.

c) Complete this sentence about these apples.

$\frac{1}{4}$ of _____ is _____ and $\frac{3}{4}$ of _____ is _____.

3 marks

Marks.......... /6

Challenge 2

1 Tara is making rectangles using different coloured squares. In each rectangle below $\frac{1}{2}$ is blue, $\frac{1}{4}$ is red and $\frac{1}{4}$ is yellow. Colour Tara's rectangles correctly.

a)

b)

c)

3 marks

Fractions of Amounts

2 Find the total of these amounts. Divide by 2 to find half, then halve the answer to find a quarter.

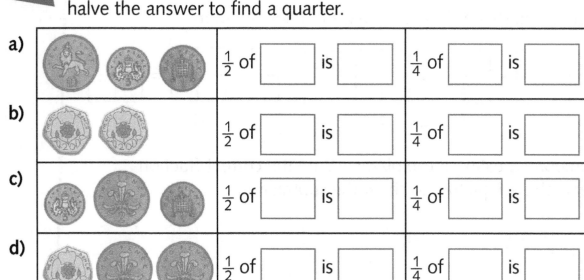

a) $\frac{1}{2}$ of ☐ is ☐ $\frac{1}{4}$ of ☐ is ☐

b) $\frac{1}{2}$ of ☐ is ☐ $\frac{1}{4}$ of ☐ is ☐

c) $\frac{1}{2}$ of ☐ is ☐ $\frac{1}{4}$ of ☐ is ☐

d) $\frac{1}{2}$ of ☐ is ☐ $\frac{1}{4}$ of ☐ is ☐

4 marks

3 Use your 2×, 3×, 4× and 5× table facts to find these fractions of amounts.

a) $\frac{1}{2}$ of 18 = ___   b) $\frac{1}{3}$ of 15 = ___   c) $\frac{1}{4}$ of 28 = ___

d) $\frac{1}{5}$ of 25 = ___   e) $\frac{1}{2}$ of 14 = ___   f) $\frac{1}{3}$ of 18 = ___

g) $\frac{1}{4}$ of 36 = ___   h) $\frac{1}{5}$ of 45 = ___

8 marks

Marks......... /15

Challenge 3

1 Put **<, >** or **=** between these statements to make them correct.

a) $\frac{2}{3}$ of 12 ☐ $\frac{1}{4}$ of 24   b) $\frac{2}{5}$ of 20 ☐ $\frac{1}{8}$ of 64

c) $\frac{2}{3}$ of 24 ☐ $\frac{2}{3}$ of 18   d) $\frac{2}{5}$ of 25 ☐ $\frac{3}{4}$ of 16

4 marks

Marks.......... /4

Total marks .... /25   How am I doing?

Addition and Subtraction of Fractions

PS Problem-solving questions

Challenge 1

1 Colour the shapes to match these fraction additions.

a) $\frac{1}{4} + \frac{1}{4} = \frac{1}{2}$

b) $\frac{1}{4} + \frac{3}{4} = 1$

2 marks

2 Someone is eating these pizzas. Write the correct fraction beneath each picture to show the subtraction.

a)

☐ − ☐ = ☐

b)
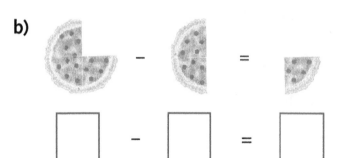

☐ − ☐ = ☐

2 marks

Marks............/4

Challenge 2

1 Add or subtract these fractions with the same denominators.

a) $\frac{1}{5} + \frac{3}{5} = $ ☐

b) $\frac{3}{10} + \frac{4}{10} = $ ☐

c) $\frac{9}{10} - \frac{6}{10} = $ ☐

d) $\frac{2}{6} + \frac{3}{6} = $ ☐

4 marks

Addition and Subtraction of Fractions

 **2** Can you find **three** different ways to make a whole by adding sixths? Colour the wheels and complete the additions below.

Example:

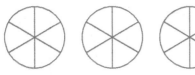

$\frac{1}{6}$ + $\frac{5}{6}$ = 1

a)

$\boxed{}$ + $\boxed{}$ = 1

b)

$\boxed{}$ + $\boxed{}$ = 1

c)

$\boxed{}$ + $\boxed{}$ = 1

 3 marks

Marks.......... /7

Challenge 3

1 Use the numbers in all three corners of the triangle to write **two** addition and **two** subtraction statements.

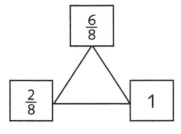

_____ + _____ = _____

_____ + _____ = _____

_____ − _____ = _____

_____ − _____ = _____

 4 marks

2 Fill in the missing fractions in these number statements.

a) $\frac{1}{8}$ + $\boxed{}$ = 1   **b)** $\boxed{}$ − $\frac{4}{10}$ = $\frac{2}{10}$ **c)** $\frac{8}{12}$ − $\boxed{}$ = $\frac{3}{12}$

3 marks

Marks.......... /7

Total marks .... /18   How am I doing?

Equivalent Fractions

1 Write the sets of equivalent fractions you can see here.

a)

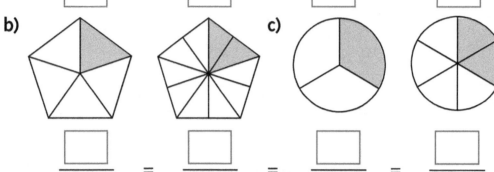

$$\frac{\square}{\square} = \frac{\square}{\square} = \frac{\square}{\square} = \frac{\square}{\square}$$

b)   c)

$$\frac{\square}{\square} = \frac{\square}{\square} = \frac{\square}{\square} = \frac{\square}{\square}$$

3 marks

2 For each pizza, colour in one half of the slices. Write the fraction that is equivalent to one half below.

a)   b)   c)   d)

$$\frac{\square}{4} \qquad \frac{\square}{8} \qquad \frac{\square}{6} \qquad \frac{\square}{10}$$

4 marks

Marks.......... /7

Use this fraction wall to help you answer the questions in Challenges 2 and 3.

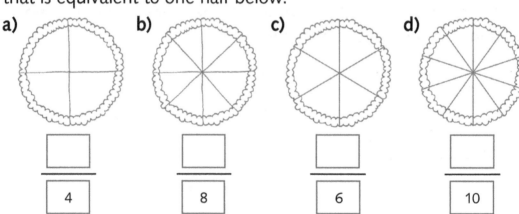

Equivalent Fractions

Challenge 2

1 Colour pairs of balloons to show the fractions that are equivalent. Use the fraction wall to help you. One has been done for you.

$\frac{2}{10}$ $\frac{4}{10}$ $\frac{1}{5}$ $\frac{2}{5}$ $\frac{4}{8}$ $\frac{2}{8}$ $\frac{1}{2}$ $\frac{3}{4}$ $\frac{6}{8}$ $\frac{1}{4}$

4 marks

2 Cross out the fraction that is NOT equivalent. Use the fraction wall to help you.

a) $\frac{1}{2}$ $\frac{5}{10}$ $\frac{6}{8}$ b) $\frac{3}{5}$ $\frac{3}{4}$ $\frac{6}{8}$

c) $\frac{2}{5}$ $\frac{3}{8}$ $\frac{4}{10}$ d) $\frac{4}{8}$ $\frac{1}{2}$ $\frac{2}{5}$

4 marks

Marks.......... /8

Challenge 3

1 Use the fraction wall to complete these equivalent fraction chains.

a) $\frac{1}{2} = \frac{2}{\square} = \frac{4}{\square}$ b) $\frac{2}{2} = \frac{4}{\square} = \frac{8}{\square}$ c) $\frac{2}{2} = \frac{5}{\square} = \frac{10}{\square}$

3 marks

2 Use doubling to complete these chains of equivalent fractions.

a) $\frac{3}{4} = \frac{6}{\square} = \frac{12}{\square}$ b) $\frac{1}{3} = \frac{2}{\square} = \frac{4}{\square}$ c) $\frac{1}{4} = \frac{2}{\square} = \frac{4}{\square}$

3 marks

3 Use halving to complete these chains of equivalent fractions.

a) $\frac{8}{16} = \frac{4}{\square} = \frac{2}{\square} = \frac{1}{\square}$ b) $\frac{8}{12} = \frac{4}{\square} = \frac{2}{\square}$ c) $\frac{20}{40} = \frac{10}{\square} = \frac{5}{\square}$

3 marks

Marks.......... /9

Total marks .... /24   How am I doing?

Comparing and Ordering Fractions

1 Write these fractions from smallest to largest.

   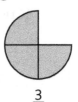

$\frac{1}{6}$ $\frac{1}{8}$ $\frac{1}{2}$ $\frac{1}{3}$ $\frac{3}{4}$ $\frac{1}{4}$

smallest ☐ ☐ ☐ ☐ ☐ ☐ **largest**

1 mark

2 Shade the shapes and then draw a ring around the larger fraction.

a) $\frac{3}{4}$ $\frac{1}{2}$

b) $\frac{1}{3}$ $\frac{4}{6}$

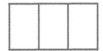 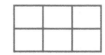

2 marks

Marks............/3

1 Write these fractions in order on the fraction number line.

$\frac{3}{8}$	$\frac{3}{4}$	$\frac{1}{4}$	$\frac{5}{8}$	$\frac{1}{8}$	$\frac{7}{8}$	$\frac{1}{2}$

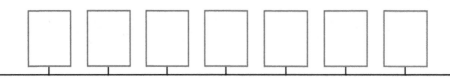

0   1

7 marks

Comparing and Ordering Fractions

2 Shade these shapes. Then choose < (less than), > (greater than) or = (equals) to compare the fractions.

a)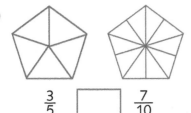

$\frac{1}{3}$ ☐ $\frac{2}{6}$

b)

$\frac{3}{4}$ ☐ $\frac{1}{2}$

c)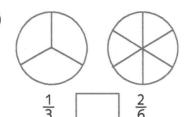

$\frac{3}{8}$ ☐ $\frac{1}{2}$

d)

$\frac{3}{5}$ ☐ $\frac{7}{10}$

e)

$\frac{1}{3}$ ☐ $\frac{2}{6}$

5 marks

Marks.........../12

Challenge 3

1 Write these fractions on this fraction number line: $\frac{1}{4}$, $\frac{1}{8}$, $\frac{1}{3}$, $\frac{1}{10}$.

Remember: the larger the denominator, the smaller the fraction.

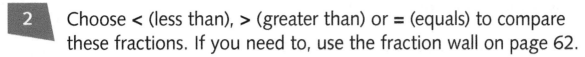

0 $\frac{1}{2}$

1 mark

2 Choose < (less than), > (greater than) or = (equals) to compare these fractions. If you need to, use the fraction wall on page 62.

a) $\frac{1}{8}$ ☐ $\frac{1}{10}$

b) $\frac{1}{4}$ ☐ $\frac{2}{8}$

c) $\frac{4}{5}$ ☐ $\frac{8}{10}$

d) $\frac{3}{4}$ ☐ $\frac{9}{10}$

e) $\frac{6}{8}$ ☐ $\frac{3}{4}$

f) $\frac{1}{4}$ ☐ $\frac{2}{5}$

6 marks

Marks.........../7

Total marks ..../22

How am I doing?

Fraction Problems

PS Problem-solving questions

Challenge 1

PS **1** 8 children in Class 3C have pets. $\frac{1}{2}$ have dogs, $\frac{1}{4}$ have cats and the rest have fish. Complete this table to show how many children have each pet.

Pet			
Number			

3 marks

PS **2** Miss Taylor is running a charity race. It is 10 km long. She has run half of the race.

a) How far has Miss Taylor run? _____ km

b) What fraction does she still have to run? _____

2 marks

Marks.......... /5

Challenge 2

PS **1** Daniel has a packet of 12 sweets.

a) Colour Daniel's sweets so that $\frac{1}{4}$ are red, $\frac{1}{4}$ are yellow, $\frac{1}{4}$ are blue and $\frac{1}{4}$ are green.

b) Daniel gives away $\frac{1}{3}$ of his sweets.
How many does he have left? _____

2 marks

PS **2** Aiden, Aaron and Ethan all have some coins.

a) Write the total amount each boy has.

Aiden	Aaron	Ethan
Total = _____	Total = _____	Total = _____

3 marks

Fraction Problems

Look at the table in part **a)**.

b) Circle true or false for each of these statements.

   i) Aaron has $\frac{1}{4}$ of the amount that Ethan has. **True / False**

  ii) Ethan has $\frac{1}{2}$ of the amount that Aiden has. **True / False**

 iii) Aaron has $\frac{1}{10}$ of the amount that Aiden has. **True / False**

 iv) Aaron has $\frac{1}{8}$ of the amount that Aiden has. **True / False**

4 marks

Marks.........../9

Challenge 3

 1 When James prepares orange squash he makes it with $\frac{1}{4}$ squash and $\frac{3}{4}$ water. Complete this table for different volumes of squash.

Orange squash	Water	Total volume
100 ml		400 ml
200 ml	600 ml	
300 ml		

4 marks

 2 **a)** Jenny is one fifth of her dad's age. Her dad is 35.
How old is Jenny? _____

b) Jenny is one quarter of her mum's age.
How old is her mum? _____

c) Jenny's dad is half Jenny's grandad's age.
How old is Jenny's grandad? _____

3 marks

Marks.........../7

Total marks ..../21   How am I doing?

1. Write **two** multiplication and **two** division statements to describe the segments in each of these chocolate bars.

a)

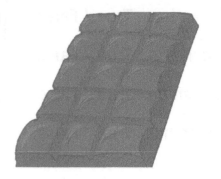

b)

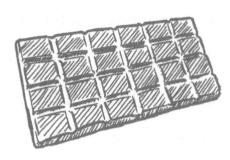

a)

_____ × _____ = _____

_____ × _____ = _____

_____ ÷ _____ = _____

_____ ÷ _____ = _____

b)

_____ × _____ = _____

_____ × _____ = _____

_____ ÷ _____ = _____

_____ ÷ _____ = _____

4 marks

2. Complete the inputs and outputs from this doubling machine.

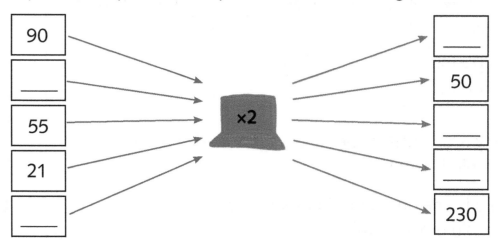

5 marks

3. Write these numbers in digits.

a) Three tens and six ones

b) Twelve tens

c) Eleven tens and three ones

d) Four hundreds, three tens and fifteen ones

4 marks

4. Use the grids to complete these multiplications.

a) 26 × 5 = ☐

×		

b) 48 × 3 = ☐

×		

5. What fraction full are these containers?

a)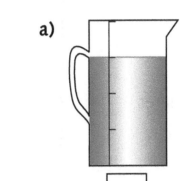

☐

b)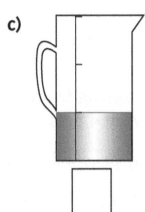

☐

c)

☐

6. Write these numbers in digits and in words.

a)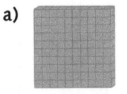

_____

_____

b)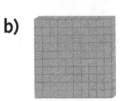

_____

_____

7. Fill in the gaps in these related multiplications.

a) 3 × 4 = 12   30 × 4 = _____   3 × 40 = _____

b) 5 × 5 = _____   50 × 5 = 250   5 × _____ = 250

c) 2 × _____ = 16   _____ × 8 = 160   2 × _____ = 160

d) _____ × 3 = 21   70 × 3 = _____   _____ × 30 = 210

8. At Osian's party there are 10 children. Divide these party items into tenths.

   a) 2000 ml ÷ 10 = _____ ml each

   b) 30 cm ÷ 10 = _____ cm each

   c) 500 g ÷ 10 = _____ g each

   d) 1000 ml ÷ 10 = _____ ml each

   4 marks

PS 9. Mrs Holt is returning felt-tipped pens to their pots. She shares 48 pens between 4 pots.

   a) How many pens will go in each pot? _____ ÷ _____ = _____

   b) Mrs Holt has found another two pots. How many will go in each pot now? _____ ÷ _____ = _____

   2 marks

10. Complete these.

   a) 3 × 4 = _____

   b) 5 × _____ = 20

   c) 48 ÷ _____ = 6

   d) _____ × 9 = 27

   4 marks

11. **3 × 5 = 15** Use this fact to complete these related divisions and multiplications.

   a) 15 ÷ 3 = _____

   b) 150 ÷ 3 = _____

   c) 5 × 3 = _____

   d) 30 × 5 = _____

   e) 150 ÷ 30 = _____

   f) 30 × 50 = _____

   6 marks

PS 12. In the library there are 30 books. $\frac{1}{5}$ are non-fiction, $\frac{1}{6}$ are dictionaries and the rest are fiction.

   How many fiction books are there? _____

   1 mark

PS **13.** Look at the number cards.

| 700 | 10 | 4 |
| 600 | 20 | 3 |

a) Choose one card from each column to make the largest number you can.

Write it in words. _____

b) Choose one card from each column to make the smallest number you can.

Write it in words. _____

14. Complete this table to show half, quarter and three quarters of each amount.

Amount	$\frac{1}{2}$	$\frac{1}{4}$	$\frac{3}{4}$
a)			
b)			
c)			
d)			

4 marks

Marks........ /47

71

Reading Scales

PS Problem-solving questions

Challenge 1

1. Use a ruler to measure the length and width of these rectangles.

   a)

   Length = _____ cm

   Width = _____ cm

   b)

   Length = _____ cm

   Width = _____ cm

   4 marks

2. How much liquid is in these containers?

   a)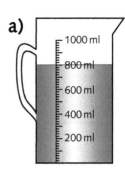
   1000 ml
   800 ml
   600 ml
   400 ml
   200 ml

   _____ ml

   b)
   1000 ml
   800 ml
   600 ml
   400 ml
   200 ml

   _____ ml

   c)
   100 ml
   80 ml
   60 ml
   40 ml
   20 ml

   _____ ml

   d)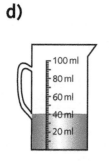
   100 ml
   80 ml
   60 ml
   40 ml
   20 ml

   _____ ml

   4 marks

3. Anil is making shortbread. He weighs out the ingredients. How much flour, butter and sugar is Anil going to use?

   a)
   500 g
   0 g   1000 g

   _____ g

   b)
   500 g
   0 g   1000 g

   _____ g

   c)
   500 g
   0 g   1000 g

   _____ g

   3 marks

Marks.......... /11

Reading Scales

Challenge 2

1 How long are these two caterpillars?

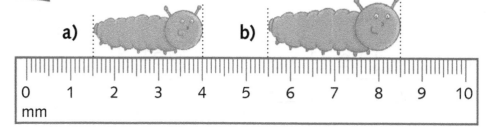

a)   b)

Caterpillar **a)** is _____ mm

Caterpillar **b)** is _____ mm

2 marks

PS **2** The temperature outside the classroom is shown on this thermometer.

The temperature inside the classroom is shown on this thermometer.

How much warmer is it inside than outside the classroom? _____ °C

1 mark

Marks.........../3

Challenge 3

PS **1** Raul has a 1 litre bottle of cola.
He pours some of it into a glass.

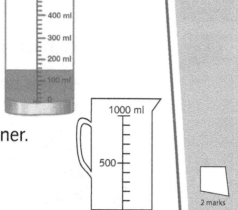

a) How much cola does Raul have left in his bottle?   _____ ml

b) Raul then empties his bottle into a new container.
Mark where his cola will come up to in this new container.

2 marks

Marks.........../2

Total marks .... /16

How am I doing?

Comparing Measures

Challenge 1

1 100 centimetres equals 1 metre.

a) Fill in the gaps in this table.

m	1	$1\frac{1}{2}$	2		3	$3\frac{1}{2}$
cm		150		250		

b) The measurements are increasing in steps of _____ cm.

6 marks

2 Ray has written his measurements in metres and centimetres, but Sal has written hers in centimetres. Draw lines to match equal lengths.

Ray
1 m 50 cm
2 m 45 cm
5 m
3 m 50 cm
5 m 8 cm

Sal
500 cm
245 cm
508 cm
150 cm
350 cm

5 marks

3 Circle the heavier weight.

a)

500 g   1 kg

b)

$\frac{1}{2}$ kg   750 g

2 marks

Marks......... /13

Challenge 2

1 Sahil, Jain, Gabriel and Ella have measured their heights.

a) Complete this table to show each of their heights in two different ways.

Sahil	127 cm	1 m 27 cm
Jain		1 m 20 cm
Gabriel	128 cm	
Ella		1 m 37 cm

b) How much taller is Ella than Jain? _____ cm

4 marks

Comparing Measures

2 Put a <, > or = sign between each pair of measurements to compare their values.

a) $1\frac{1}{2}$ litres ☐ 1500 ml   b) 2100 g ☐ $2\frac{1}{2}$ kg

c) 1 m 45 cm ☐ 138 cm   d) 22 mm ☐ 2 cm 2 mm

4 marks

Marks.......... /8

Challenge 3

PS 1 Look at this recipe for a smoothie. It serves one.

Write out the quantity of each ingredient needed for 4 people.

2 tablespoons of yogurt

60 g berries

50 ml of apple juice

_____ tablespoons of yogurt

_____ g of berries

_____ ml of apple juice

3 marks

PS 2 This model is 10 times smaller than the real helicopter.

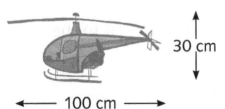

30 cm

←— 100 cm —→

What is the length and height of the real helicopter in metres?

Length = _____ m   Height = _____ m

2 marks

Marks.......... /5

Total marks .... /26   How am I doing?

Adding and Subtracting Measures

Challenge 1

PS **1** For a treat, Linda bought three bags of sweets.

 Lemon Fizzers 200 g

 Kola Kanes 150 g

 Fruit Melts 50 g

a) How many more grams of Lemon Fizzers than Fruit Melts did Linda buy? _____

b) What is the total weight of sweets that Linda bought? _____

c) Linda gave half of her sweets to her friends. How many grams was this? _____

PS **2** Supermarket Home School

←————700 m————→←——250 m——→

a) How far is it from school to the supermarket? _____ m

b) From home, how much further is it to the supermarket than to school? _____ m

 3 marks

2 marks

Marks.......... /5

Challenge 2

PS **1** Mr Smith takes a shower and uses 62 litres of water. Mrs Smith takes a bath and uses 90 litres of water.

a) How much more water does Mrs Smith use than Mr Smith? _____

b) How much water do they use altogether? _____

 2 marks

Adding and Subtracting Measures

2 Add these weights together.

a)

500 g   100 g   100 g

= _____ g

b)

500 g   20 g   20 g   20 g

= _____ g

2 marks

3 Subtract the glasses of milk from the carton.

a) 1000 ml − 400 ml = _____ ml

b) 1000 ml − 600 ml = _____ ml

2 marks

Marks.......... /6

Challenge 3

1 Hamza completes a mini-triathlon. He swims 50 m, cycles 500 m and runs 225 m.

a) What is the total distance that Hamza races? _____

b) How much further does Hamza run than swim? _____

2 marks

PS **2** How much does the banana weigh?

_____

10 g 10 g

10 g 10 g

50 g 50 g

50 g 50 g

1 mark

Marks.......... /3

Total marks .... /14   How am I doing? 😊 😐 🙁

Money

Challenge 1

1 How much money is in each purse?

a)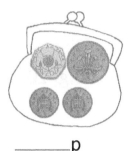

b)

c)

_____p   _____p   £_____

3 marks

PS 2 Phoebe has £10. She wants to spend some of her money.

 £3    £2    £4

How much change will Phoebe have if she buys these items?

a) 2 dolls £_____   b) 3 books £_____

c) A doll and a teddy £_____   d) 2 teddies £_____

e) Phoebe buys three items and has no change.
What could she have bought?

_____

5 marks

Marks.......... /8

Challenge 2

1 Complete the table to show the amounts in both pence and pounds. One row has been done for you.

	250p	£2.50

6 marks

PS **2** These are the prices at Jumpstart Trampoline Park.

£4.00 an hour

Additional hour £3.00

a) Bianca and her sister trampoline for an hour. How much does it cost them?

£_____

b) They pay with a £10 note. How much change do they get?

£_____

c) Rory trampolines for two hours. How much does it cost him?

£_____

d) He pays with a £20 note. How much change does he get?

£_____

 4 marks

Marks.........../10

Challenge 3

1 Complete these calculations.

a)
```
    £ 6 . 6 5
  + £ 2 . 6 5
  _____

  _____
```

b)
```
    £ 5 . 1 6
  − £ 1 . 3 6
  _____

  _____
```

c)
```
    £ 9 . 4 0
  − £ 4 . 1 1
  _____

  _____
```

d)
```
    £ 8 . 0 6
  + £ 7 . 0 7
  _____

  _____
```

 4 marks

2 Evan has £2.00 in 20p and 10p pieces. He has three times as many 10p pieces as 20p pieces. How many of each coin does Evan have?

 10p

20p

 2 marks

Marks........../6

Total marks ..../24   How am I doing?   😣

Time

Challenge 1

1 Draw the hour and minute hands to show these times.

a) Twenty past 6 **b)** Quarter past 1 **c)** Ten to 4

3 marks

2 Write these times in words

a)

_____

b)

_____

c)

_____

d)

_____

4 marks

Marks............/7

Challenge 2

1 Draw lines to match the analogue clocks to the digital clocks that show the same time.

07:07   11:57   08:08   03:20   08:35

5 marks

2 What numbers do these Roman numerals represent?

VI	IV	XI	III	IX	VIII

6 marks

Marks.......... /11

Challenge 3

1 Write the digital times shown on these analogue clocks.

a)    **b)**   **c)**   **d)**

[__ : __]   [__ : __]   [__ : __]   [__ : __]

4 marks

 2 12:21 is a palindromic time. It's the same read forwards or backwards.

Find two other 4-digit palindromic times.

__ __ : __ __

__ __ : __ __

2 marks

Marks.......... /6

Total marks .... /24   How am I doing?

81

Units of Time

Challenge 1

1 Fill in the gaps in this rhyme.

'Thirty days has _____, April, June and

_____. All the rest have

_____ except _____

alone which has _____ clear, twenty-nine
in each leap year.'

5 marks

2 **a)** How many seconds in one minute? ☐

b) How many minutes in one hour? ☐

c) How many hours in one day? ☐

d) How many days in two weeks? ☐

4 marks

Marks.........../9

Challenge 2

1 Fill in the gaps in this table.

Seconds	15		45	60		
Minutes		$\frac{1}{2}$		1	$1\frac{1}{2}$	2

5 marks

2 Write <, > or = to make these time statements correct.

a) 120 seconds ☐ 2 minutes   **b)** 21 days ☐ 4 weeks

c) 5 weeks ☐ 35 days   **d)** A fortnight ☐ 12 days

4 marks

Units of Time

3

a) How many days are there altogether in July and August?

b) 2040 will be a leap year. How many days will there be altogether in February and March 2040?

2 marks

4

a) How many days in 2 weeks and 2 days?

b) How many minutes in 1 hour and 20 minutes?

c) How many seconds in 5 minutes?

d) How many hours in two days?

4 marks

Marks......... /15

Challenge 3

1 A year is a leap year if it can be divided by 4 without a remainder. Circle the leap years.

2002   2004   2009   2012   2014   2016   2018   2020

4 marks

2 Put these measures of time in order from shortest to longest.

15 sec   1 min   95 sec $1\frac{1}{2}$ min   59 sec   95 min $\frac{1}{2}$ min

shortest _____ _____ _____ _____

_____ _____ _____ longest

7 marks

Marks......... /11

Total marks .... /35   How am I doing?   ☺ 😐 ☹

Duration of Events

Challenge 1

1 How long is it between these pairs of times?

a)

b)

_____   _____

2 marks

PS **2** Eva is allowed to watch an hour and a half of TV. She starts watching at 4 o'clock. What time does she finish?

_____

1 mark

PS **3** Fyn's guitar lesson is 30 minutes long. It finishes at 7:30 p.m.

What time did it start? _____

1 mark

Marks.......... /4

Challenge 2

1 At Crescent Park Primary, this is the school day.

School starts	Lunchtime begins	Lunchtime ends	School ends

a) How long is the morning session? _____

b) How long is the lunch break? _____

c) How long is the afternoon session? _____

d) How long is the school day? _____

4 marks

Duration of Events

PS | **2** | Abi's train journey takes 45 minutes. The train leaves at quarter to 3. What time should Abi arrive?

_____

1 mark

PS | **3** | Umair swims for 50 minutes. His lesson ended at 5:30 p.m. What time did it start?

_____

1 mark

Marks.......... /6

Challenge 3

PS | **1** | After school, Jamie is watching TV. This is the programme guide for when Jamie is watching.

4:30   5:00   5:30

Kidz	Toons		Dance Club	Magix
News	World Weather	News	Sports	
World	Reptile Zone	Animal Antics		Ocean View

Fill in the gaps.

He watches Toons until Animal Antics starts at _____.

He watches Animal Antics, which lasts _____ minutes.

Finally he watches Magix for _____ minutes until 5:30 p.m.

Altogether, Jamie has watched TV for _____ minutes.

4 marks

Marks.......... /4

Total marks .... /14   How am I doing?

85

Perimeter of Shapes

PS Problem-solving questions

Challenge 1

1 Use a ruler to measure the perimeter (the distance all the way around) of each rectangle.

a)

______ cm

b)

______ cm

c)

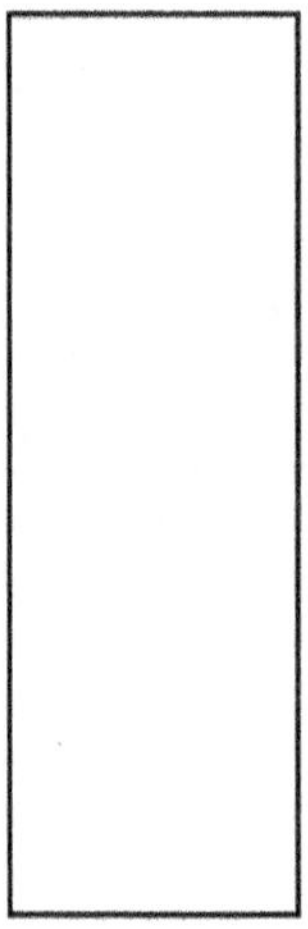

______ cm

d)

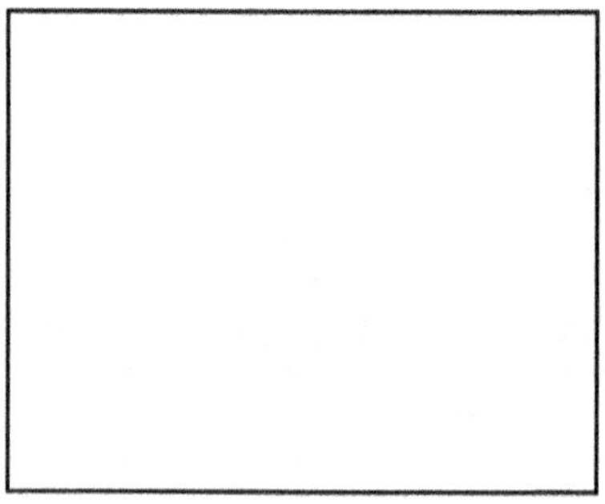

______ cm

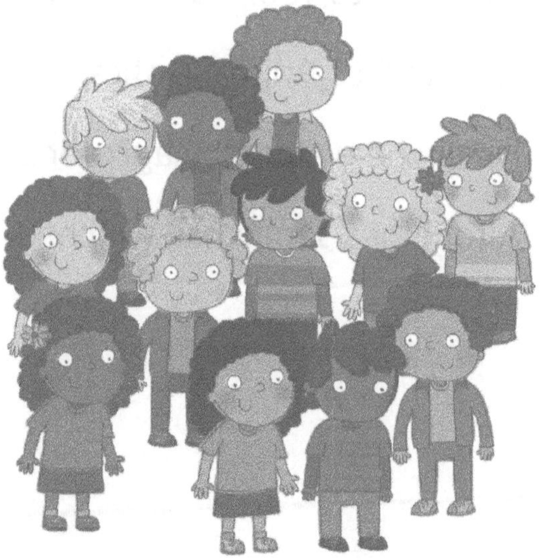

4 marks

Marks........ /4

Perimeter of Shapes

Challenge 2

1 Use a ruler to measure the perimeter of each of these regular polygons.

a)

b)

c)

d)

4 marks

Marks.......... /4

Challenge 3

PS **1** This rectangle has a perimeter of 24 cm.

10 cm

2 cm

Find four different rectangles that all have a perimeter of 24 cm.

Width	Length

4 marks

Marks.......... /4

Total marks .... /12   How am I doing?   😊   😐   😣

Progress Test 3

1. Put these sets of measures in order from smallest to largest.

   a) 1 m 55 cm   140 cm $1\frac{1}{2}$ m

   _____   _____   _____

   b) 1300 ml   1 litre 250 ml $1\frac{1}{2}$ litres

   _____   _____   _____

   c) 46 mm   4 cm 4 mm $4\frac{1}{2}$ cm

   _____   _____   _____

   d) 2 kg   3000 g   2 kg 500 g

   _____   _____   _____

   4 marks

2. Colour half of each shape and write the equivalent fraction below.

   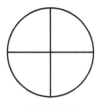 

   4 marks

3. Write the time shown on these analogue clocks.

   a)    b)

   _____   _____

   c)   d)

   _____   _____

   4 marks

PS **4.** What is the smallest number of coins that you could use to pay for these amounts? Draw the coins in the table.

£1.06	
83p	
19p	
44p	

4 marks

PS **5.** From Monday to Friday, Reception class receive 30 pieces of fruit. Circle the calculation that would tell you how many pieces of fruit they receive in a week.

$$30 + 30 = 60 \qquad 30 \div 5 = 6 \qquad 5 \times 30 = 150$$

$$5 + 30 = 35 \qquad 30 \times 7 = 210$$

1 mark

PS **6.** Gus goes to bed at 8 o'clock and wakes up at half past 7. His older sister goes to bed at half past 9 and wakes up at 8 o'clock.

How much more sleep does Gus get than his sister? _____

1 mark

PS **7.** Draw a rectangle that has a perimeter of 12 cm.

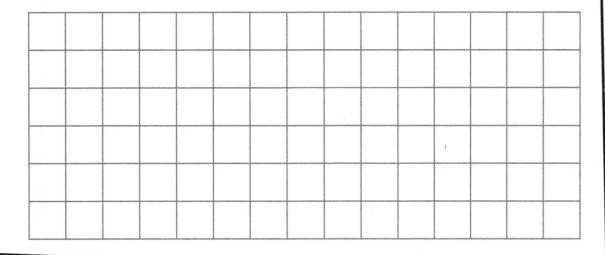

1 mark

8. Fill in the boxes around each number to show what is 10 and 100 more and less.

a)

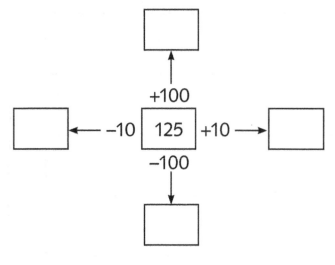

b)

2 marks

9. Use a ruler to measure these lines.

a) _____ _____ cm

b) _____ _____ cm

c) _____ _____ cm

d) _____ _____ cm

4 marks

10. Complete these sentences.

a) _____ is the inverse of multiplication.

b) Addition is the inverse of _____.

2 marks

11. Look at the liquid in container A.

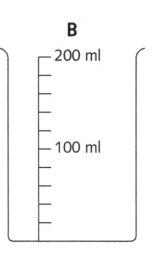

A

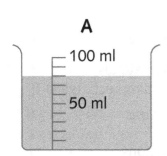

Rachel pours all the liquid into container B. Mark where it will come up to.

1 mark

12. Complete these calculations.

a)
```
  £ 3 . 3 5
– £ 1 . 3 0
_____
```

b)
```
  £ 1 . 0 5
+ £ 7 . 0 8
_____
```

c)
```
  £ 8 . 4 4
+ £ 7 . 2 4
_____
```

d)
```
  £ 6 . 4 2
– £ 3 . 0 6
_____
```

4 marks

PS **13.** Louis measured his shadow three times during the day. At 9:00 a.m. it measured 175 cm. At 12:15 p.m. it had become 55 cm shorter. At 3:00 p.m. it was 30 cm longer than it was at 12:15 p.m. Complete Louis' table of results.

Time	9:00 a.m.	12:15 p.m.	3:00 p.m.
Shadow length			

3 marks

Marks........ /35

2-D Shapes

Challenge 1

1 Use four different colours to show the different polygons that make up this tractor. Colour the key to show which colour you have used for each shape.

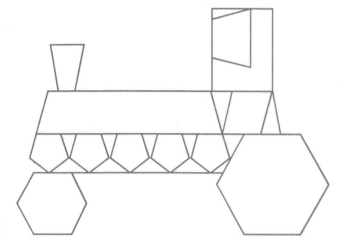

KEY

☐ Rectangles

☐ Quadrilaterals (excluding rectangles)

☐ Hexagons

☐ Pentagons

4 marks

2 Circle **true** or **false** for each of these statements.

a) A square has 4 equal sides.   **True / False**

b) A rectangle has 2 pairs of parallel sides.   **True / False**

c) A pentagon always has equal sides.   **True / False**

d) A hexagon has 5 sides.   **True / False**

e) A square is a quadrilateral.   **True / False**

5 marks

Marks.......... /9

Challenge 2

1 How many sides are there altogether on these shapes?

a) 6 triangles ☐   b) 5 pentagons ☐

c) 9 triangles ☐   d) 10 hexagons ☐

4 marks

2-D Shapes

2 Draw a rectangle that measures 2 cm by 5 cm.

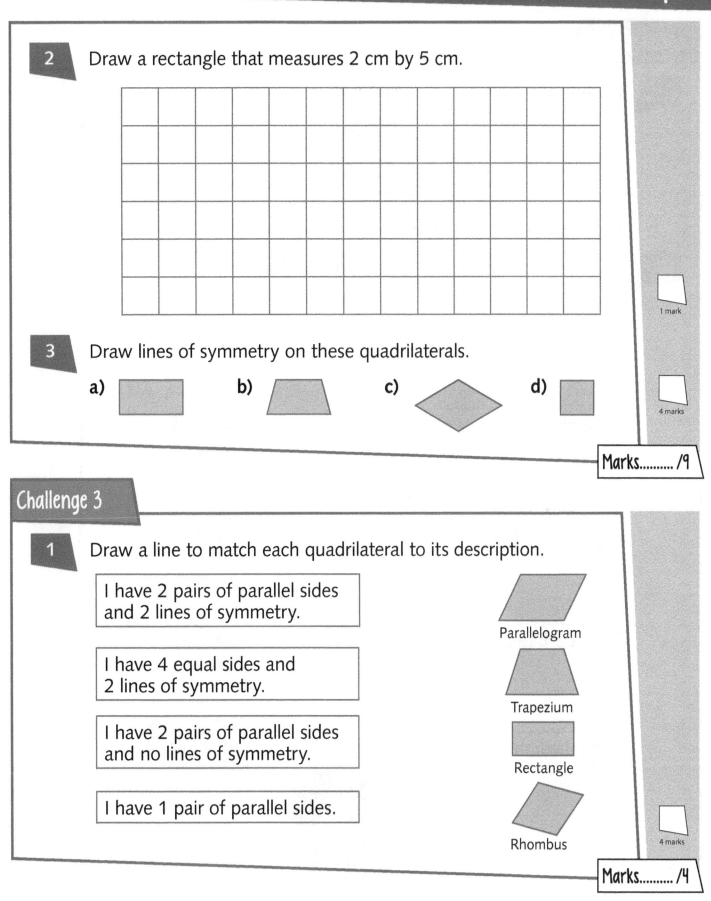

1 mark

3 Draw lines of symmetry on these quadrilaterals.

a)   b)   c)   d)

4 marks

Marks.......... /9

Challenge 3

1 Draw a line to match each quadrilateral to its description.

I have 2 pairs of parallel sides and 2 lines of symmetry.

I have 4 equal sides and 2 lines of symmetry.

I have 2 pairs of parallel sides and no lines of symmetry.

I have 1 pair of parallel sides.

Parallelogram

Trapezium

Rectangle

Rhombus

4 marks

Marks.......... /4

Total marks .... /22   How am I doing?

3-D Shapes

Challenge I

1 **a)** Add the labels 'Face', 'Edge' and 'Vertex' to this cube.

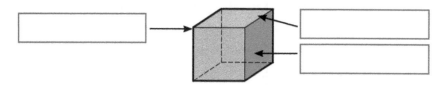

b) A cube has _____ faces, _____ edges and _____ vertices.

6 marks

2 Draw lines to match these 3-D shapes to their names.

| Triangular prism |

| Cuboid |

| Hexagonal prism |

| Square-based pyramid |

| Tetrahedron |

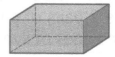

5 marks

Marks......... /11

3-D Shapes

Challenge 2

1 Name these 3-D shapes from the shapes of their faces. One has been done for you.

Cube				

4 marks

2 What am I?

a) I have 6 faces that are all square. _____

b) I have 6 edges and all my faces are the same.

_____

c) I have 6 vertices. _____

d) I have an odd number of vertices. _____

4 marks

Marks............/8

Challenge 3

1 Claudette is trying to build a cube using her construction kit. Tick the shapes that would fold up to make a cube.

A   B   C   D

2 marks

Marks.........../2

Total marks ..../21   How am I doing?

95

Lines

PS **1** **a)** Use two colours to trace the vertical and horizontal lines in this rectangle.

b) Now use the same colours on all the vertical and horizontal lines you can see in this flag.

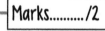

2 marks

Marks............/2

1 Lines that are always the same distance apart are parallel. Colour pairs of parallel lines in these parallelograms the same colour.

 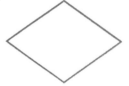 

3 marks

2 Lines that meet at a right angle are perpendicular. Colour pairs of perpendicular lines in these rectangles the same colour.

3 marks

Marks.......... /6

Challenge 3

1 In this rectangle (shape 1), line AB is parallel to line CD and perpendicular to lines AC and BD.

Look at this octagon (shape 2).

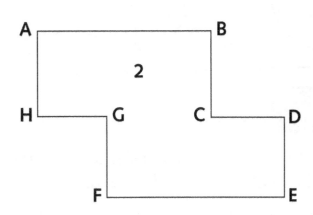

Complete these sentences.

a) One of the lines parallel to line AB is line _____.

b) One of the lines perpendicular to line AH is line _____.

c) One of the lines parallel to line CD is line _____.

d) One of the lines perpendicular to line FE is line _____.

4 marks

2 Use different colours to show parallel and perpendicular lines on this Union Jack.

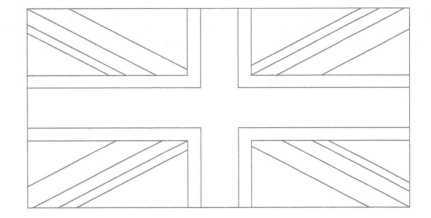

2 marks

Marks.......... /6

Total marks .... /14   How am I doing?

Right Angles

Make a right angle tester from a circle of paper. Fold it in half and then into quarters to form a right angle.

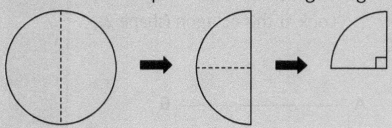

Challenge 1

1 Use your right angle tester to label the right angles in these flags. One has been found for you.

a)

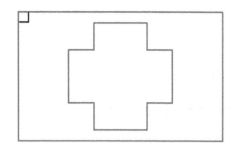

b)

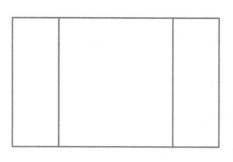

6 marks

Marks.......... /6

Challenge 2

1 Every triangle has three angles. Use two colours to show the angles that are greater than a right angle and those that are less than a right angle. Use your right angle tester to help you.

a)   b)   c)

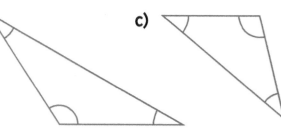

KEY

☐ Less than a right angle

☐ Greater than a right angle

3 marks

Marks.......... /3

Right Angles

Challenge 3

1 Use a ruler to draw these shapes.

   a) A right-angled triangle

   b) A pentagon with two right angles

2 marks

2 Draw lines to match each quadrilateral to its description.

I have 4 right angles.

Parallelogram

My opposite angles are equal.

Trapezium

I have 1 pair of parallel sides.

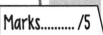

Rectangle

3 marks

Marks.......... /5

Total marks .... /14   How am I doing?

Angles and Turns

Challenge 1

1 Niamh is facing north.
In which direction will she be facing after the following turns?

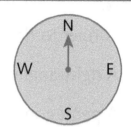

a) One right angle clockwise _____

b) One right angle anticlockwise _____

c) Two right angles anticlockwise _____

d) Three right angles clockwise _____

4 marks

2 How many right angles will Niamh need for these turns?

a) From east to west

☐ right angles clockwise or

☐ right angles anticlockwise

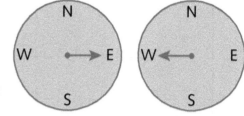

b) From south to east

☐ right angles clockwise or

☐ right angles anticlockwise

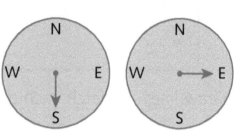

2 marks

Marks.......... /6

Challenge 2

1 The minute hand can only turn clockwise around these clock faces. Draw its position after one right angle turn.

a)  **b)**

2 marks

Angles and Turns

2 How many right angles has the minute hand turned clockwise between these positions?

a)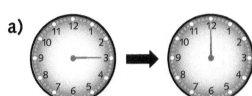

_____

b)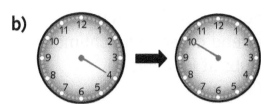

_____

2 marks

Marks.......... /4

Challenge 3

1 Sakina's remote-controlled car can make a $\frac{1}{4}$ turn left (lt90), a $\frac{1}{4}$ right (rt90) or move forwards a certain number of squares, for example 3 squares (fd3).

Write the program to get Sakina's car to the finish square.

a) _____   b) _____   c) _____   d) _____   e) _____

f) _____   g) _____   h) _____   i) _____

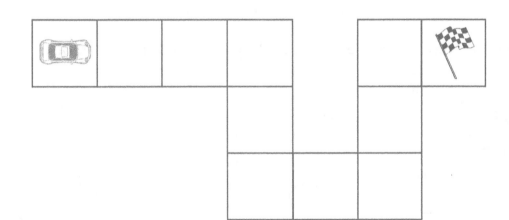

4 marks

Marks.......... /4

Total marks .... /14   How am I doing?

Bar Charts

1 This bar chart shows the number of children who attend different lunchtime clubs at Hazelwood Primary.

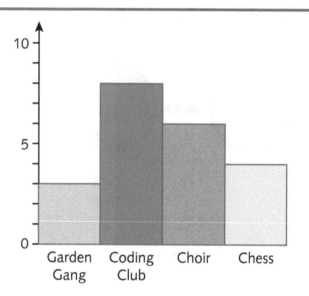

a) Which club is the most popular? _____

b) Which club is the least popular? _____

c) How many children attend Coding Club? _____

d) How many more children attend Choir than Garden Gang? _____

e) How many children attend lunchtime clubs? _____

5 marks

Marks.......... /5

1 The children were asked about clubs they attended after school.

Club	Number
Football	26
Swimming	14
Dance	9
Drama	20
Tennis	13

Draw bars to complete this bar graph.

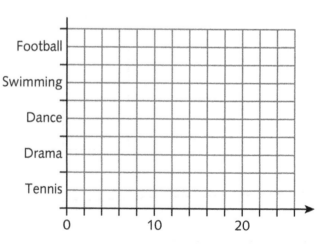

5 marks

Marks.......... /5

Bar Charts

Challenge 3

1 All the children at Hazelwood Primary were asked how they travel to school.

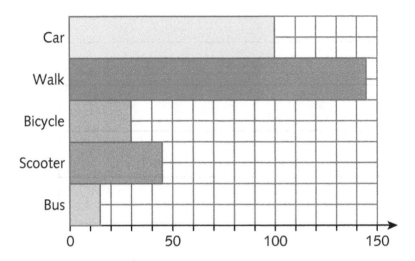

a) How many children walk to school? _____

b) How many children take the bus? _____

c) How many more children come on a scooter than on a bicycle? _____

d) How many children are there at Hazelwood Primary? _____

e) The headteacher says: 'More than twice as many children walk, cycle or come on a scooter as come by car.' Is he right? Explain how you know.

_____

_____

5 marks

Marks.......... /5

Total marks .... /15   **How am I doing?**

Pictograms

Challenge 1

PS **1** Class 5 collected shells on their residential trip to the seaside. They made a pictogram where one picture represents every two shells that they found.

Cockle	
Grey top	
Mussel	
Periwinkle	

a) How many periwinkle shells did Class 5 find? _____

b) Which shell was the most common? _____

c) Which shell was twice as common as mussels? _____

d) How many shells did Class 5 find altogether? _____

4 marks

Marks.......... /4

Challenge 2

1 After their trip to the beach the children had to choose a sandwich for the following day.

Cheese	Tuna	Egg	Chicken
6	7	5	12

Complete the pictogram to show their sandwich choices.

Key = 2 sandwiches   = 1 sandwich

Pictograms

Cheese	
Tuna	
Egg	
Chicken	

4 marks

Marks.......... /4

Challenge 3

PS **1** The children bought souvenirs in the Sea Life Museum. This pictogram shows their purchases.

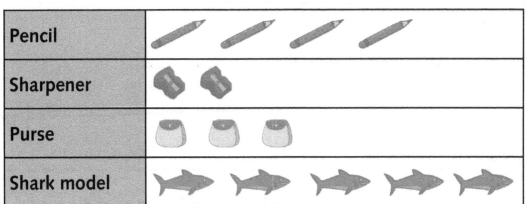

Pencil	
Sharpener	
Purse	
Shark model	

a) The children bought 25 model sharks. How many purchases does one picture represent in the pictogram? _____

b) How many pencils did they buy? _____

c) In the shop, purses cost £2.00 each. How much did the children spend on purses? _____

d) Sharpeners cost 85p each. How much did the children spend on sharpeners? _____

4 marks

Marks.......... /4

Total marks .... /12

How am I doing?

Tables

Challenge 1

1

1. At Outdoor World, they sell four different tents.

Tent	Colour	Sleeps	Price
Valley	green	6	£150
Wilderness	green	2	£100
Festival	red	2	£70
Glen	navy blue	8	£200

a) How much more expensive is the Wilderness than the Festival? _____

b) How many more people can sleep in the Glen than the Valley? _____

c) How many different colours of tent are there? _____

d) Which green tent is cheaper? _____

4 marks

Marks.......... /4

Challenge 2

1

The store manager made a tally of tents sold in May and June.

Tent	May	June	Total
Valley	││││	├┼┼ ├┼┼ ││	_____
Wilderness	├┼┼ │││	├┼┼ │││	16
Festival	├┼┼	├┼┼ ├┼┼ ├┼┼ ├┼┼	_____
Glen	├┼┼ ││	├┼┼ ││	_____
Total	24	_____	_____

a) Complete the table by adding the missing totals.

5 marks

b) In which month were the most tents sold? _____

How many tents was this? _____

2 marks

c) Which model of tent sold the most? _____

How many tents was this? _____

2 marks

d) Which two models of tent sold the same amount?

_____ and _____

2 marks

Marks.......... /11

Challenge 3

1 Outdoor World also sells hiking boots. This table shows the sizes of all the boots sold at the weekend.

$7\frac{1}{2}$	12	8	8	9	4
9	8	11	10	$10\frac{1}{2}$	$4\frac{1}{2}$
$8\frac{1}{2}$	$6\frac{1}{2}$	9	$11\frac{1}{2}$	10	5
10	$8\frac{1}{2}$	$7\frac{1}{2}$	10	9	$6\frac{1}{2}$
5	5	6	$7\frac{1}{2}$	6	$4\frac{1}{2}$

a) Put the information into this frequency table. One group has been done for you.

Size	$3\frac{1}{2}$–5	$5\frac{1}{2}$–7	$7\frac{1}{2}$–9	$9\frac{1}{2}$–11	$11\frac{1}{2}$–13
Number	6				

4 marks

b) Which was the most popular size range of hiking boot?

_____

1 mark

Marks.......... /5

Total marks .... /20   How am I doing?

1. This pictogram shows the number of tablet computers in each class at Woodland Academy.

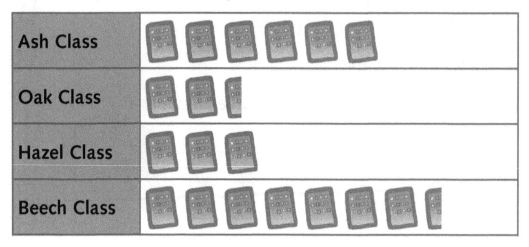

Ash Class	
Oak Class	
Hazel Class	
Beech Class	

There are 12 tablets in Ash Class.

a) How many tablets does 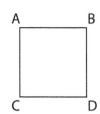 represent? _____

b) How many tablets are there in Oak Class? _____

c) How many more tablets are there in Beech Class than in Ash Class? _____

3 marks

2. Complete these calculations.

a) $1 - \frac{3}{5} = \boxed{}$

b) $\frac{2}{10} + \frac{1}{10} = \boxed{}$

c) $\frac{6}{8} - \frac{2}{8} = \boxed{}$

d) $\frac{2}{3} + \frac{1}{3} = \boxed{}$

e) $1 - \frac{1}{4} = \boxed{}$

f) $\frac{2}{10} + \frac{3}{10} + \frac{4}{10} = \boxed{}$

6 marks

3. Look at this square.

A ____ B
| |
| |
C   D

a) Colour all the horizontal lines one colour and all the vertical lines another colour.

1 mark

b) Make these statements correct by crossing out the word that is NOT correct.

    i) Line AB is **parallel** / **perpendicular** to line BD.

    ii) Line AB is **parallel** / **perpendicular** to line CD.

2 marks

4. Add lines of symmetry to these 2-D shapes.

a)

b)

c)

d)

4 marks

5. Write these fractions in order from smallest to largest.

$\frac{1}{5}$	$\frac{1}{3}$	$\frac{1}{4}$	$\frac{1}{10}$	$\frac{1}{8}$	$\frac{1}{2}$

1 mark

6. A remote-controlled car is placed on 1 cm square paper. It can make a $\frac{1}{4}$ turn left (lt90), a $\frac{1}{4}$ right (rt90) or move forwards a certain number of centimetres, for example 3 centimetres (fd3).

Write a program to make the car draw a square with sides of 6 cm.

_____ _____ _____ _____

_____ _____ _____

2 marks

109

7. Janek has been counting the birds visiting his bird table. He has recorded the number of each bird as a tally mark.

Bird	Tally	Number
Chaffinch	⦀⦀ ⦀⦀ ‖	12
Starling	⦀⦀ ⦀⦀ ⦀⦀ ⦀⦀ ⦀⦀	
Blackbird	⦀⦀ ‖	
Robin	‖‖	

a) Complete Janek's table.

3 marks

b) Use Janek's data to draw a bar chart to display his information.

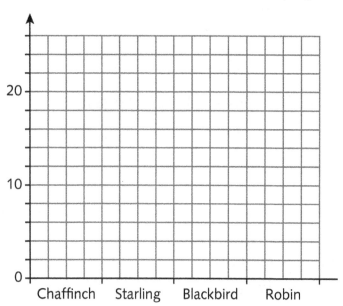

4 marks

c) How many more starlings than robins did Janek count? _____

1 mark

d) How many birds did Janek count altogether? _____

1 mark

8. Write two multiplication and two subtraction statements to match this array of cupcakes.

_____ × _____ = _____   _____ ÷ _____ = _____

_____ × _____ = _____   _____ ÷ _____ = _____

4 marks

9. Complete the table below.

A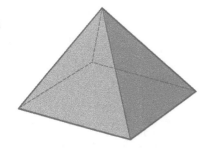

B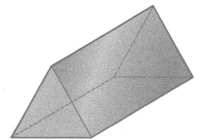

	Name of shape	Number of faces	Number of edges	Number of vertices
A				
B				

8 marks

10. Write **<**, **>** or **=** between these pairs of calculations to make each number statement correct.

a) 8 × 5 ☐ 5 × 8

b) £1.50 + £1.00 ☐ £3.00 − £1.50

c) 45 ÷ 5 ☐ 70 ÷ 10

d) 198 + 3 ☐ 205 − 7

4 marks

11. Mark the right angles in these shapes.

a)

b)

c)

d)

4 marks

12. Write these times in order from shortest to longest.

a fortnight	a week	10 seconds	12 hours
an hour	a day	30 minutes	

shortest _____

_____ longest

1 mark

Marks........ /49

111

Pages 4–11
Starter Test

1. a) 16   b) 61
   c) 108   d) 50
2. 5, 15, 46, 51, 55, 68, 92
3. a) 8, 14
   b) 15, 6
   c) 25, 30, 45
   d) 80, 60, 50
4. a)

		14	
23	24	25	
		34	

   b)

		57	
66	67	68	
		77	

   c)

		35	
44	45	46	
		55	

   d)

		9	
18	19	20	
		29	

5. a) 37   b) 71
   c) 43p
6. a) <   b) >
   c) <   d) =
7. 16 + 4 = 20   20 − 4 = 16
   5 + 15 = 20   20 − 15 = 5
   14 + 6 = 20   20 − 6 = 14
8. a) 8   b) 4
   c) 6   d) 9
   e) 27   f) 20
9. a) 19   b) 25 cm
10. 3 × 4 = 12 and 4 × 3 = 12
11. a) 18
    b) 9
    c) 40
    d) 100
    e) 21
    f) 10
12. a) 20
    b) 10
    c) 2
    d) 4
13. a) 12
    b) 15
    c) 12

14.

Number halved	Number	Number doubled
50	100	12
10	6	16
3	8	200
4	12	24
6	20	40

15. a) 8
    b) 4
    c) 12
16. $1\frac{1}{4}$, $1\frac{3}{4}$, $2\frac{1}{2}$
17. a) 3 cm, 2 cm
    b) 6 cm, 1 cm
    c) 5 cm, 4 cm
18. a) 800 g
    b) 150 cm
    c) 45°C
    d) 250 ml
19. a)    b)

    c)

20. a) 25p
    b) Four different answers where the coins all total 25p, e.g. 10p, 10p, 2p, 2p and 1p.
21. a) 30p, 70p
    b) Four different possibilities that total 50p, e.g. two drinks and a cupcake.
22. 1 mark for each type of shape coloured correctly, up to a total of 4.

Answers

23.

Shape	Faces	Edges	Vertices
(cube)	6	12	8
(pyramid)	5	8	5
(triangular prism)	5	9	6

24. Three clock faces showing the minute hand at:
  a) 6
  b) 3
  c) 9
25. a) 5
  b) 10
  c) $3\frac{1}{2}$ cars drawn in the Silver row.

Pages 12–13
1. a) 67 = 60 + 7
  b) 19 = 10 + 9
  c) 88 = 80 + 8
  d) 55 = 50 + 5
  e) 24 = 20 + 4
  f) 70 = 70 + 0
2. a) 4
  b) 10
  c) 90
  d) 60
3. a) 62
  b) 19
  c) 34
  d) 48
Challenge 2
1. a) 419 = 400 + 10 + 9
  b) 229 = 200 + 20 + 9
  c) 305 = 300 + 0 + 5
  d) 980 = 900 + 80 + 0
2. a) 700
  b) 5
  c) 70
  c) 900
Challenge 3
1. a) 844
  b) 504
  c) 260
  d) 483

2. a) =   b) <
  c) >   d) =

Pages 14–15
Challenge 1
1. a) 37
  b) 125
  c) 250
  d) 302
2. a) 58p   b) 75p
Challenge 2
1. a) 120
  b) 150
  c) 190
  d) 610
  e) 630
  f) 680
2. a) £2.12
  b) £3.03
Challenge 3
1. Accept values (inclusive) in these ranges:
  a) 23–27
  b) 43–47
  c) 85–89
2. a) 20p, 20p, 2p, 2p
  b) 50p, 20p, 20p, 5p, 2p, 2p
  c) £2, £1, 5p, 2p, 1p

Pages 16–17
Challenge 1
1. a) 73
  b) 17
  c) 44
  d) 200
  e) 606
  f) 158
2. a) Thirty-five
  b) Eleven
  c) Four hundred and fifty
  d) Two hundred and seven
Challenge 2
1. a) 598
  b) Five hundred and ninety-eight
2. a) 413
  b) 182
Challenge 3
1. a) 8741
  b) Eight thousand seven hundred and forty-one
  c) 1478
  d) One thousand four hundred and seventy eight
2. Three pounds and eighty-eight pence

Pages 18–19
Challenge 1
1. a) 78p   b) 29p
    c) 83p
2. a) 17p   b) 41p
    c) 34p
3. a)

41	**42**
	52
	62

    b)

	79
88	89
98	

    c)

11	**12**
21	22

    d)

14	
24	25
	26

Challenge 2
1.

10 less	Start number	10 more
60	70	**80**
20	30	40
15	25	**35**
45	**55**	65
5	**15**	**25**

2. a) 595 kg, 795 kg
    b) 210 kg, 410 kg
Challenge 3
1. a) 285   b) 311
    c) 23

Pages 20–21
Challenge 1
1. a) 8, 20
    b) 24, 48
    c) 100, 200, 250
    d) 300, 400, 600
        1 mark for each completed rocket.

Challenge 2
1. a) 8
    b) 50
    c) 4
    d) 4
2. a) 500, 400, 300
    b) 28, 32, 36
    c) 72, 64, 56
    d) 1000, 1050, 1100
    e) 100, 104, 108
    f) 40, 48, 56
Challenge 3
1. Multiples of 4   68, 72, 76, 80, 84, 88, 92
    Multiples of 8   56, 64, 72, 80, 88, 96, 104

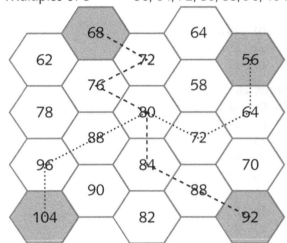

2. Many examples possible, e.g.
    a) 100
    b) 400

Pages 22–23
Challenge 1
1. a) 106, 116, 160, 161, 166
2. a) 199 to 201 inclusive
    b) 246 to 253 inclusive
3. a) 99, 100, 101, 105
    b) 385, 386, 388, 390, 391, 392
Challenge 2
1. 2, 5, 1, 4, 3
    1 mark for all correct.
2. 1, 5, 3, 2, 4
    1 mark for all correct.
3. a) <   b) >
    c) <   d) >
Challenge 3
1. a) =   b) <
    c) >   d) >
2. 259, 295, 529, 592, 925, 952
    1 mark for each correct number and 1 further
    mark for the correct order.

Answers

Pages 24–25
Challenge 1
1. Caamir 45 cm, Blanka 25 cm
2. Joe 50 m, Liu 250 m
3. a) 6
   b) 600
Challenge 2
1. a) 98 cm
   b) 118 cm
2. a) 261
   b) 61
3. a) 250
   b) 4
Challenge 3
1. a) 1999
   b) 1909
2. a) £1.50
   b) £3.50

Pages 26–27
Challenge 1
1. a) 49
   b) 21
   c) 131
   d) 240
   e) 204
   f) 825
2. a) 14
   b) 69
   c) 339
   d) 94
   e) 548
   f) 753
Challenge 2
1. a) 145
   b) 318
   c) 831
2. a) 180
   b) 175
   c) 493
Challenge 3
1. a) 267 + 252 = (200 + 200) + (60 + 50) + (7 + 2) = 400 + 110 + 9 = 519
   b) 514 + 388 = (500 + 300) + (10 + 80) + (4 + 8) = 800 + 90 + 12 = 902
2. a) 79, +2, +70, +7
   b) 67, +8, +50, +9

Pages 28–29
Challenge 1
1. a) 80   b) 25
   c) 95   d) 45
   e) 65   f) 0

2. a) 72
   b) 49
   c) 21
   d) 56
   e) 93
   f) 68
Challenge 2
1. a) 31p
   b) 78p
   c) 27p
   d) 19p
2. a) 63p
   b) 44p
   c) 81p
   d) 28p
Challenge 3
1. a) i) £1.68 ii) £1.33 iii) £1.49 iv) £1.16
   b) Sweets and a drink (ii and iv)
   c) £2.66

Pages 30–31
Challenge 1
1. a) 9, 70, 79
   b) 10, 80, 90
   c) 12, 80, 92
   d) 9, 80, 89
Challenge 2
1. a) 88
   b) 63
   c) 90
   d) 79
2. a) 288
   b) 567
   c) 948
   d) 782
Challenge 3
1. a) 345 + 224
   b) 402 + 238
   c) 68 + 47
   d) 460 + 170
   e) 215 + 719

Pages 32–33
Challenge 1
1. a) 44
   b) 43
   c) 924
   d) 125
   e) 321
   f) 362
   g) 330
   h) 443

116

Challenge 2

1. a) 321
    b) 119
    c) 244
    d) 818
2. a) **56** – 35
    b) **89** – **12**
    c) **3**50 – **30**
    d) 944 – **732**
    e) 263 – 145

Challenge 3

1. a) 482
    b) 361
    c) 766
    d) 158

Pages 34–35
Challenge 1

1. a) 60
    b) 500
    c) 120 cm
    d) 70
    e) 5 kg
2. a) 19 + 6 = 25
    b) 130 – 65 = 65
    c) 45p + 55p = £1.00

Challenge 2

1. a) 150
    b) 300
    c) 500
    d) 4
2. a) 11 + 24 = 35
    b) 140 ÷ 10 = 14 or 140 ÷ 14 = 10
    c) 4 × 30 = 120
    d) 880 – 310 = 570 or 880 – 570 = 310

Challenge 3

1. a) Accept answers from 13–17 inclusive
    b) Accept answers from 35–45 inclusive
    c) Accept answers from 55–65 inclusive

Pages 36–37
Challenge 1

1. a) 140
    b) 95
    c) 360
2. a) 150
    b) 570

Challenge 2

1. a) 134 cm
    b) 155 cm
    c) 21 cm

2. a) 69
    b) 15
3. £36
4. a)

6	1	8
7	5	3
2	9	4

   b)

6	7	2
1	5	9
8	3	4

   c)

8	1	6
3	5	7
4	9	2

Challenge 3

1. a) 2p + 10p + 10p OR 1p + 1p + 20p
    b) £1.65

2.

		84		
	41		43	
15		26		17
1	14		12	5

Pages 38–41
Progress Test 1

1. 489, 498, 505, 515, 551
2. a) 53
    b) 87
    c) 41
    d) 26
3. a) 982
    b) 837
    c) 899
    d) 718
4. 7 cm
5. a)

+	75	**85**
40	**115**	125
55	**130**	**140**

   b)

+	95	65
80	**175**	**145**
65	160	130

6. a) 17
    b) 22
    c) 19
    d) 152
7. a) 600 + 70 + 2
    b) 100 + 20 + 3
8. a) 620   b) 650
    c) 695
9. 32p

Answers

10. a) 97   **b)** 104
     c) 315   **d)** 355
11. a) 40, 44, 48
     b) 56, 48, 40
12. a) 199   **b)** 639
13. a) 50   **b)** 8
     c) 200
14. a) £305   **b)** £195

Pages 42–43
Challenge 1
1. **a)** $3 \times 8 = 24$ $8 \times 3 = 24$ (1 mark)
     $24 \div 8 = 3$ $24 \div 3 = 8$ (1 mark)
    b) $7 \times 4 = 28$ $4 \times 7 = 28$ (1 mark)
     $28 \div 7 = 4$ $28 \div 4 = 7$ (1 mark)

2. **a)**

×	3	10	2
5	15	50	10
2	6	20	4
4	12	40	8

    b)

×	3	4	5
2	6	8	10
8	24	32	40
3	9	12	15

Challenge 2
1. **a)** $6 \times 8 = 48$, $8 \times 6 = 48$ (1 mark)
     $48 \div 8 = 6$, $48 \div 6 = 8$ (1 mark)
    b) $9 \times 3 = 27$, $3 \times 9 = 27$ (1 mark)
     $27 \div 9 = 3$, $27 \div 3 = 9$ (1 mark)
    c) $12 \times 8 = 96$, $8 \times 12 = 96$ (1 mark)
     $96 \div 8 = 12$, $96 \div 12 = 8$ (1 mark)

Challenge 3
1. **a)**

×	8	3	4
2	16	6	8
8	64	24	32
5	40	15	20

    b)

×	4	2	3
8	32	16	24
10	40	20	30
4	16	8	12

2. **a)** 80
    b) 300
    c) 640
    d) 20
    e) 80
    f) 40
    g) 400
    h) 900

Pages 44–45
Challenge 1
1. 45 – 90; 11 – 22; 65 – 130; 30 – 60; 70 – 140; 25 – 50
2. **a)** 40
    b) 60
    c) 35
    d) 22
    e) 160
    f) 24

Challenge 2
1. **a)** 14, **28**
    b) **24**, 48
    c) 75, **150**
    d) **120**, 240
    e) 90, **180**
    f) 160, **320**
2. **a)** Bilal £2.40   **b)** Zain £1.20
    c) Akila £4.80

Challenge 3
1. $8 = \mathbf{1} \times 8 = \mathbf{2} \times 4 = \mathbf{4} \times 2$
    $16 = \mathbf{2} \times 8 = \mathbf{4} \times 4 = \mathbf{8} \times 2$
    $24 = \mathbf{3} \times 8 = \mathbf{6} \times 4 = \mathbf{12} \times 2$
2. 900 metres
3. £2.00

Pages 46–47
Challenge 1
1.

	×1	×2	×3	×4	×5	×6	×7	×8	×9	×10	×11	×12
3	3	6	9	12	15	18	21	24	27	30	33	36
4	4	8	12	16	20	24	28	32	36	40	44	48
8	8	16	24	32	40	48	56	64	72	80	88	96

1 mark for each correct row.

2. **a)** 16
    b) 40
    c) 18
    d) 9
    e) 24
    f) 56
3. 32
4. 28

Challenge 2
1. **a)** 3   **b)** 4
    c) 96   **d)** 7
    e) 64   **f)** 4
2. **a)** £48
    b) £27
    c) 5 tennis racquets
    d) 10

Challenge 3
1.

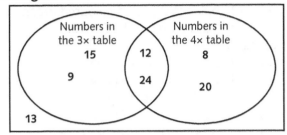

Numbers in the 3× table: 15, 9
12, 24 (overlap)
Numbers in the 4× table: 8, 20
13

2 marks for all 7 numbers placed in the correct zones. 1 mark for 5 or 6 numbers placed in the correct zones.

Pages 48–49
Challenge 1
1. These pairs of bricks in the same colour:
   $8 \times 5 \times 2 = 80$, $5 \times 8 \times 2 = 80$
   $10 \times 3 \times 2 = 60$, $2 \times 3 \times 10 = 60$
   $4 \times 3 \times 2 = 24$, $2 \times 4 \times 3 = 24$
   $5 \times 2 \times 4 = 40$, $4 \times 5 \times 2 = 40$
   2 marks for each correct pair of multiplications.
2. a) >   b) =
   c) <   d) =

Challenge 2
1. a) $5 \times 30 = $ **150**, $50 \times 3 = $ **150**
   b) $240 \div 4 = $ **60**, $240 \div 40 = $ **6**
   c) **40** $= 80 \div 2$, **4** $= 80 \div 20$
   d) $40 \times 3 = $ **120**, $4 \times 30 = $ **120**
   1 mark for each correct train.
2. a) <   b) <
   c) <   d) =

Challenge 3
1. a) 96
   b) $2 \times 5 \times 16 = 10 \times 16 = 160$
   c) $3 \times 2 \times 9 = 6 \times 9 = 54$
2. Variety of answers possible, e.g.
   a) $80 \times 4 = 320$ and $8 \times 40 = 320$
   b) $360 \div 12 = 30$ and $360 \div 120 = 3$
   c) $70 \times 4 = 280$ and $7 \times 40 = 280$

Pages 50–51
Challenge 1
1. a) 230   b) 120
   c) 510   d) 880
   e) 490   f) 170
2. a) $20 \times 5 + 5 \times 5 = 100 + 25 = 125$
   b) $10 \times 4 + 8 \times 4 = 40 + 32 = 72$
   c) $40 \times 3 + 2 \times 3 = 120 + 6 = 126$

Challenge 2
1. a) 216

×	20	7
8	160	56

b) 192

×	30	2
6	180	12

2. 120
3. 174

Challenge 3
1. a) 710

×	100	40	2
5	500	200	10

b) 924

×	200	30	1
4	800	120	4

2. £7.00
3. £1.50

Pages 52–53
Challenge 1
1. a) 30   b) 60
   c) 5
2. a) 2   b) 4
   c) 3

Challenge 2
1. a) $8 \times 4 = 32$
   b) $120p \div 40p = 3$
   c) $80p \times 2 = £1.60$
2. 4 biscuits
3. 8 rooms

Challenge 3
1. a) 60
   b) 120
   c) 210
   d) 7
2. a) 11
   b) 14

Pages 54–55
Challenge 1
1. a) Any 5 sections coloured, e.g.

b) Any 3 sections coloured, e.g.

Answers

c) Any 6 sections coloured, e.g.

2. a) 4 coloured green, 1 orange, 2 yellow and 3 red
b) $\frac{3}{10}$

Challenge 2
1. $\frac{3}{10}, \frac{6}{10}, \frac{7}{10}, \frac{9}{10}$
2. a) 2   **b)** 7
    c) 10p   **d)** 3 cm
    e) 8 kg   **f)** 10 km

Challenge 3
1. 40
2. a) £40   **b)** 65 m
    c) 8   **d)** 5
    e) 33   **f)** 70
3. a) 12   **b)** 27
    c) $\frac{3}{10}$

Pages 56–57
Challenge 1
1. a) $\frac{1}{5}$
    b) $\frac{1}{6}$
2. a) $\frac{1}{4}$
    b) $\frac{1}{2}$
    c) $\frac{3}{4}$
3. Coloured to show 6 beads yellow, 3 red and 3 blue.

Challenge 2
1. a) $\frac{3}{10}$ **b)** $\frac{4}{6}$ or $\frac{2}{3}$
2. a) $\frac{2}{8}$ or $\frac{1}{4}$ **b)** $\frac{4}{8}$ or $\frac{1}{2}$
    c) $\frac{6}{8}$ or $\frac{3}{4}$
3. a) $\frac{7}{10}$ **b)** $\frac{3}{10}$

Challenge 3
1. a) 2 more hexagons should be coloured.
    b) 3 more squares should be coloured.
2. a) $\frac{3}{4}$
    b) $1\frac{1}{4}$
    c) $1\frac{1}{2}$
3. $\frac{1}{4}$

Pages 58–59
Challenge 1
1. a) 4 planes circled and coloured orange.
    b) 2 planes circled and coloured blue.
    c) $\frac{1}{2}$ of **8** is **4** and $\frac{1}{4}$ of **8** is **2**.

2. a) 4 apples circled and coloured red.
    b) 12 apples circled and coloured green.
    c) $\frac{1}{4}$ of **16** is **4** and $\frac{3}{4}$ of **16** is **12**.

Challenge 2
1. a) 4 blue, 2 red and 2 yellow
    b) 2 blue, 1 red and 1 yellow
    c) 6 blue, 3 red and 3 yellow
2. a) $\frac{1}{2}$ of 16p is 8p, $\frac{1}{4}$ of 16p is 4p
    b) $\frac{1}{2}$ of 40p is 20p, $\frac{1}{4}$ of 40p is 10p
    c) $\frac{1}{2}$ of 8p is 4p, $\frac{1}{4}$ of 8p is 2p
    d) $\frac{1}{2}$ of 24p is 12p, $\frac{1}{4}$ of 24p is 6p
3. a) 9   **b)** 5
    c) 7   **d)** 5
    e) 7   **f)** 6
    g) 9   **h)** 9

Challenge 3
1. a) >   **b)** =
    c) >   **d)** <

Pages 60–61
Challenge 1
1. Squares coloured appropriately.
2. a) $1 - \frac{1}{4} = \frac{3}{4}$ **b)** $\frac{3}{4} - \frac{1}{2} = \frac{1}{4}$

Challenge 2
1. a) $\frac{4}{5}$ **b)** $\frac{7}{10}$
    c) $\frac{3}{10}$ **d)** $\frac{5}{6}$
2. 1 mark for each correct solution, e.g.
$\frac{3}{6} + \frac{3}{6} = 1$

Challenge 3
1. $\frac{6}{8} + \frac{2}{8} = 1$, $\frac{2}{8} + \frac{6}{8} = 1$, $1 - \frac{2}{8} = \frac{6}{8}$, $1 - \frac{6}{8} = \frac{2}{8}$
2. a) $\frac{7}{8}$ **b)** $\frac{6}{10}$
    c) $\frac{5}{12}$

Pages 62–63
Challenge 1
1. a) $\frac{1}{2} = \frac{2}{4} = \frac{3}{6} = \frac{4}{8}$
    b) $\frac{1}{5} = \frac{2}{10}$
    c) $\frac{1}{3} = \frac{2}{6}$
2. a) $\frac{2}{4}$ **b)** $\frac{4}{8}$
    c) $\frac{3}{6}$ **d)** $\frac{5}{10}$
1 mark for each correct fraction and coloured pizza.

Challenge 2
1. 1 mark for each pair of fractions coloured as follows: $\frac{1}{2}$ and $\frac{4}{8}$; $\frac{2}{5}$ and $\frac{4}{10}$; $\frac{1}{4}$ and $\frac{2}{8}$; $\frac{3}{4}$ and $\frac{6}{8}$
2. a) $\frac{6}{8}$ **b)** $\frac{3}{5}$
    c) $\frac{3}{8}$ **d)** $\frac{2}{5}$

Challenge 3

1. a) $\frac{1}{2} = \frac{2}{4} = \frac{4}{8}$ b) $\frac{2}{2} = \frac{4}{4} = \frac{8}{8}$

 c) $\frac{2}{2} = \frac{5}{5} = \frac{10}{10}$

2. a) $\frac{3}{4} = \frac{6}{8} = \frac{12}{16}$ b) $\frac{1}{3} = \frac{2}{6} = \frac{4}{12}$

 c) $\frac{1}{4} = \frac{2}{8} = \frac{4}{16}$

3. a) $\frac{8}{16} = \frac{4}{8} = \frac{2}{4} = \frac{1}{2}$ b) $\frac{8}{12} = \frac{4}{6} = \frac{2}{3}$

 c) $\frac{20}{40} = \frac{10}{20} = \frac{5}{10}$

Pages 64–65
Challenge 1

1. $\frac{1}{8}, \frac{1}{6}, \frac{1}{4}, \frac{1}{3}, \frac{1}{2}, \frac{3}{4}$
2. a) Shapes shaded correctly; $\frac{3}{4}$ circled.
   b) Shapes shaded correctly; $\frac{4}{6}$ circled.

Challenge 2

1. $\frac{1}{8}, \frac{1}{4}, \frac{3}{8}, \frac{1}{2}, \frac{5}{8}, \frac{3}{4}, \frac{7}{8}$
2. a) =   b) >
   c) <   d) <
   e) =

Challenge 3

1. $\frac{1}{10}, \frac{1}{8}, \frac{1}{4}, \frac{1}{3}$
2. a) >   b) =
   c) =   d) <
   e) =   f) <

Pages 66–67
Challenge 1

1. Dogs 4, Cats 2, Fish 2
2. a) 5 km   b) $\frac{1}{2}$

Challenge 2

1. a) 3 coloured red, 3 yellow, 3 blue and 3 green.
   b) 8
2. a) Aiden 40p, Aaron 5p, Ethan 20p
   b) i) True   ii) True
      iii) False   iv) True

Challenge 3

1.

Orange squash	Water	Total volume
100 ml	**300 ml**	400 ml
200 ml	600 ml	**800 ml**
300 ml	**900 ml**	1200 ml

2. a) 7   b) 28
   c) 70

Pages 68–71
Progress Test 2

1. a) 3 × 5 = 15, 5 × 3 = 15, 15 ÷ 5 = 3, 15 ÷ 3 = 5
   b) 4 × 6 = 24, 6 × 4 = 24, 24 ÷ 6 = 4, 24 ÷ 4 = 6
   1 mark for each correct pair of calculations.

2. 90, **180**; **25**, 50; 55, **110**; 21, **42**; **115**, 230
3. a) 36   b) 120
   c) 113   d) 445
4. a) 130

×	20	6
5	100	30

 b) 144

×	40	8
3	120	24

5. a) $\frac{3}{4}$ b) $\frac{4}{6}$
   c) $\frac{1}{3}$
6. a) 101; one hundred and one
   b) 111; one hundred and eleven
7. a) 120, 120   b) 25, 50
   c) 8, 20, 80   d) 7, 210, 7
   1 mark for each correct set of calculations.
8. a) 200 ml   b) 3 cm
   c) 50 g   d) 100 ml
9. a) 48 ÷ 4 = 12   b) 48 ÷ 6 = 8
10. a) 12   b) 4
    c) 8   d) 3
11. a) 5   b) 50
    c) 15   d) 150
    e) 5   f) 1500
12. 19
13. a) Seven hundred and twenty-four
    b) Six hundred and thirteen
14.

Amount	$\frac{1}{2}$	$\frac{1}{4}$	$\frac{3}{4}$
(coins)	10p	5p	15p
(coins)	12p	6p	18p
(coins)	£1.00	50p	£1.50
(coins)	62p	31p	93p

Pages 72–73
Challenge 1

1. a) Length = 8 cm, Width = 2 cm
   b) Length = 6 cm, Width = 3 cm
2. a) 800 ml   b) 200 ml
   c) 90 ml   d) 40 ml
3. a) 300 g   b) 400 g
   c) 200 g

Answers

Challenge 2

1. a) 25 mm   b) 30 mm
2. 14°C

Challenge 3

1. a) 850 ml
   b) Indication of liquid midway between 800 ml and 900 ml.

Pages 74–75

Challenge 1

1. a)

m	1	1½	2	2½	3	3½
cm	100	150	200	250	300	350

   b) 50 cm
2. 1 m 50 cm – 150 cm; 2 m 45 cm – 245 cm; 5 m – 500 cm; 3 m 50 cm – 350 cm; 5 m 8 cm – 508 cm
3. a) 1 kg circled   b) 750 g circled

Challenge 2

1. a)

Sahil	127 cm	1 m 27 cm
Jain	120 cm	1 m 20 cm
Gabriel	128 cm	**1 m 28 cm**
Ella	**137 cm**	1m 37 m

   b) 17 cm
2. a) =   b) <
   c) >   d) =

Challenge 3

1. **8 tablespoons** of yogurt, **240 g** of berries, **200 ml** of apple juice
2. Length = 10 m   Height = 3 m

Pages 76–77

Challenge 1

1. a) 150 g
   b) 400 g
   c) 200g
2. a) 950 m
   b) 450 m

Challenge 2

1. a) 28 litres
   b) 152 litres
2. a) 700 g
   b) 560 g
3. a) 600 ml   b) 400 ml

Challenge 3

1. a) 775 m   b) 175 m
2. 160 g

Pages 78–79

Challenge 1

1. a) 24p   b) 65p
   c) £1.06

2. a) £4   b) £4
   c) £3   d) £2
   e) 2 dolls and 1 teddy or 2 teddies and a book

Challenge 2

1.

	250p	£2.50
	320p	£3.20
	155p	£1.55
	204p	£2.04

2. a) £8.00   b) £2.00
   c) £7.00   d) £13

Challenge 3

1. a) £9.30   b) £3.80
   c) £5.29   d) £15.13
2. 12 × 10p and 4 × 20p

Pages 80–81

Challenge 1

1. a)   b)   c)

2. a) Five past three
   b) Twenty-five to eleven
   c) Twenty-five past two
   d) Ten past seven

Challenge 2

1.

| 07:07 | 11:57 | 08:08 | 3:20 | 08:35 |

2.

VI	IV	XI	III	IX	VIII
6	4	11	3	9	8

Challenge 3

1. a) 05:38
   b) 02:18
   c) 12:19
   d) 02:47

2. There are a number of examples, e.g. 05:50 or 11:11.

Pages 82–83
Challenge 1
1. September, November, 31, February, 28
2. **a)** 60   **b)** 60
   c) 24   **d)** 14

Challenge 2
1.

Seconds	15	30	45	60	90	120
Minutes	$\frac{1}{4}$	$\frac{1}{2}$	$\frac{3}{4}$	1	$1\frac{1}{2}$	2

2. **a)** =   **b)** <
   c) =   **d)** >
3. **a)** 62
   b) 60
4. **a)** 16   **b)** 80
   c) 300   **d)** 48

Challenge 3
1. 2004, 2012, 2016 and 2020 circled.
2. 15 sec, $\frac{1}{2}$ min, 59 sec, 1 min, $1\frac{1}{2}$ min, 95 sec, 95 min

Pages 84–85
Challenge 1
1. **a)** 2 hrs
   b) $3\frac{1}{2}$ hrs
2. Half past 5
3. 7:00 p.m.

Challenge 2
1. **a)** 3 hrs and 10 mins
   b) 50 mins
   c) 2 hrs and 20 mins
   d) 6 hrs and 20 mins
2. Half past 3
3. 4:40 p.m.

Challenge 3
1. 4:45, 30 mins, 15 mins, 1 hour

Pages 86–87
Challenge 1
1. **a)** 16 cm   **b)** 14 cm
   c) 16 cm   **d)** 18 cm

Challenge 2
1. **a)** 9 cm   **b)** 12 cm
   c) 10 cm   **d)** 16 cm

Challenge 3
1. Answers with whole centimetre sides are 9 cm by 3 cm, 8 cm by 4 cm, 7 cm by 5 cm and 6 cm by 6 cm.

Pages 88–91
Progress Test 3
1. **a)** 140 cm, $1\frac{1}{2}$ m, 1 m 55 cm
   b) 1 litre 250 ml, 1300 ml, $1\frac{1}{2}$ litres
   c) 4 cm 4 mm, $4\frac{1}{2}$ cm, 46 mm
   d) 2 kg, 2 kg 500 g, 3000 g
2. $\frac{2}{4}$, $\frac{4}{8}$, $\frac{8}{16}$, $\frac{5}{10}$
3. **a)** Twelve minutes past 7
   b) Twenty-five to 12
   c) Ten past 8
   d) Twenty-eight minutes to 1
4.

£1.06	£1, 5p, 1p
83p	50p, 20p, 10p 2p, 1p
19p	10p, 5p, 2p, 2p
44p	20p, 20p, 2p, 2p

5. 5 × 30 = 150
6. 1 hour
7. Variety of possible answers, e.g. 5 cm × 1 cm or 4 cm × 2 cm
8. **a)** 225, 115, 135, 25
   b) 593, 483, 503, 393
9. **a)** 10 cm   **b)** 7 cm
   c) 4 cm   **d)** 5 cm
10. **a)** Division   **b)** Subtraction
11. Liquid indicated at 80 ml mark.
12. **a)** £2.05   **b)** £8.13
   c) £15.68   **d)** £3.36
13.

Time	9:00 a.m.	12:15 p.m.	3:00 p.m.
Shadow length	175 cm	120 cm	150 cm

Pages 92–93
Challenge 1
1. 1 mark for each of the named shapes correctly coloured and marked in the key.
2. **a)** True   **b)** True
   c) False   **d)** False
   e) True

Challenge 2
1. **a)** 18   **b)** 25
   c) 27   **d)** 60
2. Shape drawn with sides within 2 mm of given dimensions.
3. **a)**   **b)**

   c) 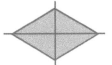   **d)**

Answers

Challenge 3
1.

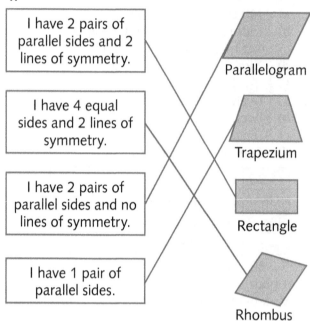

I have 2 pairs of parallel sides and 2 lines of symmetry.

I have 4 equal sides and 2 lines of symmetry.

I have 2 pairs of parallel sides and no lines of symmetry.

I have 1 pair of parallel sides.

Parallelogram

Trapezium

Rectangle

Rhombus

Pages 94–95
Challenge 1
1. a)

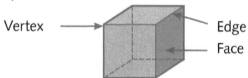

Vertex

Edge

Face

   b) A cube has **6** faces, **12** edges and **8** vertices.
2.

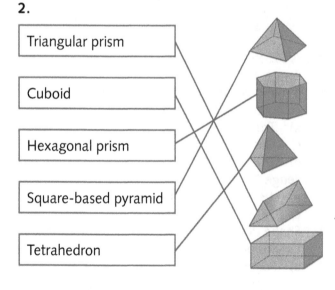

Triangular prism

Cuboid

Hexagonal prism

Square-based pyramid

Tetrahedron

Challenge 2
1.

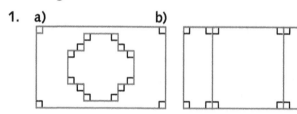			
Hexagonal prism	Square-based pyramid	Cuboid	Tetra-hedron

2. a) Cube
   b) Tetrahedron
   c) Triangular prism
   d) Square-based pyramid

Challenge 3
1. Shapes A and D ticked

Pages 96–97
Challenge 1
1. a) and b) 1 mark for the lines on each shape coloured appropriately.

Challenge 2
1. 1 mark for each correctly coloured shape.
2. 1 mark for each correctly coloured shape.

Challenge 3
1. a) HG, CD or FE   b) AB or HG
    c) AB or FE   d) GF or DE
2. 1 mark for all parallel lines and 1 mark for all perpendicular lines indicated correctly.

Pages 98–99
Challenge 1
1. a)   b)

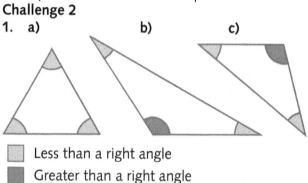

    Up to 3 marks for each shape.

Challenge 2
1. a)   b)   c)

    ☐ Less than a right angle
    ■ Greater than a right angle

Challenge 3
1. a) For example:

    b) For example:

2.

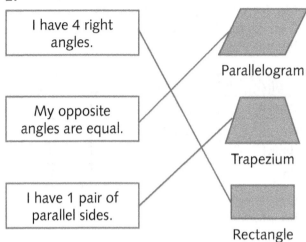

I have 4 right angles.	Parallelogram
My opposite angles are equal.	Trapezium
I have 1 pair of parallel sides.	Rectangle

Pages 100–101
Challenge 1
1. **a)** East
   b) West
   c) South
   d) West
2. **a)** **2** right angles clockwise or **2** right angles anticlockwise
   b) **3** right angles clockwise or **1** right angle anticlockwise

Challenge 2
1. **a)**

b)

2. **a)** 3   **b)** 2

Challenge 3
1. **a)** fd3   **b)** rt90
   c) fd2   **d)** lt90
   e) fd2   **f)** lt90
   g) fd2   **h)** rt90
   i) fd1

Maximum of 4 marks; reduce by 1 mark for each mistake or omission.

Pages 102–103
Challenge 1
1. **a)** Coding Club   **b)** Garden Gang
   c) 8   **d)** 3
   e) 21

Challenge 2
1.

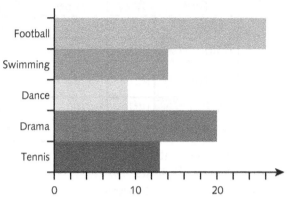

1 mark for each bar drawn correctly.

Challenge 3
1. **a)** 145
   b) 15
   c) 15
   d) 335
   e) Yes because 220 children walk, cycle or come on a scooter. This is more than double the number who come by car, which is 100.

Pages 104–105
Challenge 1
1. **a)** 6
   b) Grey top
   c) Cockle
   d) 28

Challenge 2
1.

Cheese	🧀 🧀 🧀
Tuna	🧀 🧀 🧀 🧀
Egg	🧀 🧀 🧀
Chicken	🧀 🧀 🧀 🧀 🧀 🧀

1 mark for each row drawn correctly.

Challenge 3
1. **a)** 5
   b) 20
   c) £30
   d) £8.50

Pages 106–107
Challenge 1
1. **a)** £30
   b) 2
   c) 3
   d) Wilderness

Answers

Challenge 2

1. a)

Tent	May	June	Total
Valley	IIII	LHT LHT II	**16**
Wilderness	LHT IIII	LHT III	16
Festival	LHT	LHT LHT LHT LHT	25
Glen	LHT II	LHT II	14
Total	24	**47**	**71**

b) June; 47 tents
c) Festival; 25 tents
d) Valley and Wilderness

Challenge 3

1. a)

Size	$3\frac{1}{2}$ –5	$5\frac{1}{2}$ –7	$7\frac{1}{2}$ –9	$9\frac{1}{2}$ –11	$11\frac{1}{2}$ –13
Number	6	4	12	6	2

b) $7\frac{1}{2}$ –9

Pages 108–111
Progress Test 4

1. a) 2   b) 5
   c) 3

2. a) $\frac{2}{5}$ b) $\frac{3}{10}$
   c) $\frac{4}{8}$ d) $\frac{3}{3}$ or 1
   e) $\frac{3}{4}$ f) $\frac{9}{10}$

3. a) Vertical and horizontal lines coloured differently.
   b) i) Line AB is perpendicular to line BD.
      ii) Line AB is parallel to line CD.

4. a) 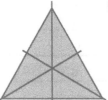   b)

   c) 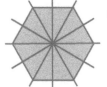   d)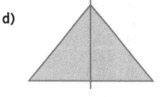

5. $\frac{1}{10}, \frac{1}{8}, \frac{1}{5}, \frac{1}{4}, \frac{1}{3}, \frac{1}{2}$
   1 mark for all in the correct order.

6. lt90, fd6, rt90, fd6, rt90, fd6, rt90, fd6 or fd6, lt90, fd6, lt90, fd6, lt90, fd6
   2 marks for the correct program, and 1 mark if there is a single mistake or omission.

7.

Bird	Tally	Number
Chaffinch	LHT LHT II	12
Starling	LHT LHT LHT LHT LHT	25
Blackbird	LHT II	7
Robin	III	3

b) Use Janek's data to draw a bar chart to display his information.

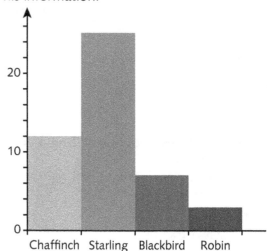

c) 22   d) 47

8. (In any order) 4 × 3 = 12, 12 ÷ 3 = 4, 3 × 4 = 12, 12 ÷ 4 = 3

9.

	Name of shape	Number of faces	Number of edges	Number of vertices
A	Square-based pyramid	5	8	5
B	Triangular prism	5	9	6

10. a) =   b) >
    c) >   d) >

11. a)   b)   c)   d)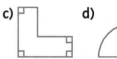

12. 10 seconds, 30 minutes, an hour, 12 hours, a day, a week, a fortnight
    1 mark for all in the correct order.

Progress Test 1

Q	Topic	✓ or ✗	See page
1	Comparing and Ordering Numbers		22
2	Number Bonds		28
3	Adding in Columns		30
4	Addition and Subtraction Problems		36
5	Adding and Subtracting Mentally		26
6	Subtracting in Columns		32
7	Place Value		12
8	Representing Numbers		14
9	Number Problems		24
10	Reading and Writing Numbers		16
11	Counting in Multiples		20
12	Addition and Subtraction Problems		36
13	Estimating and Checking Calculations		34
14	10 and 100 More or Less		18

Progress Test 2

Q	Topic	✓ or ✗	See page
1	Multiplication and Division Facts		42
2	Doubling and Halving		44
3	Place Value		12
4	Multiplying 2-digit Numbers		50
5	Recognising Fractions		56
6	Representing Numbers		14
7	Mental Multiplication and Division		48
8	Tenths		54
9	Mental Multiplication and Division		48
10	3×, 4× and 8× Tables		46
11	Mental Multiplication and Division		48
12	Fraction Problems		66
13	Reading and Writing Numbers/Place Value		16/12
14	Fractions of Amounts		58

Progress Test Charts

Progress Test 3

Q	Topic	✓ or ✗	See page
1	Comparing Measures		74
2	Equivalent Fractions		62
3	Time		80
4	Money		78
5	Multiplication and Division Problems		52
6	Duration of Events		84
7	Perimeter of Shapes		86
8	10 and 100 More or Less		18
9	Reading Scales		72
10	Estimating and Checking Calculations		34
11	Reading Scales		72
12	Money/Adding in Columns/Subtracting in Columns		78/30/32
13	Adding and Subtracting Measures		76

Progress Test 4

Q	Topic	✓ or ✗	See page
1	Pictograms		104
2	Addition and Subtraction of Fractions		60
3	Lines		96
4	2-D Shapes		92
5	Comparing and Ordering Fractions		64
6	Angles and Turns		100
7	Bar Charts and Tables		102
8	Multiplication and Division Facts		42
9	3-D Shapes		94
10	Comparing and Ordering Numbers		22
11	Right Angles		98
12	Units of Time		82

What am I doing well in? _____

_____

What do I need to improve? _____

_____